Muhammad Afzal Upal

Moderate Fundamentalists

Muhammad Afzal Upal

Moderate Fundamentalists

The Ahmadiyya Muslim Jama'at in the Lens of Cognitive Science of Religion

Managing Editor: Katarzyna Tempczyk

Associate Editor: Rasa Pranskevičiūtė

Language Editor: Wayne Smith

DE GRUYTER

ISBN: 978-3-11-055648-3
e-ISBN: 978-3-11-055664-3

Library of Congress Cataloging-in-Publication Data
A CIP catalog record for this book has been applied for at the Library of Congress.

Published by De Gruyter Open Ltd, Warsaw/Berlin
Part of Walter de Gruyter GmbH, Berlin/Boston
The book is published with open access at www.degruyter.com.

Managing Editor: Katarzyna Tempczyk
Associate Editor: Rasa Pranskevičiūtė
Language Editor: Wayne Smith

www.degruyteropen.com
Cover illustration: © Thinkstock

Contents

Acknowledgement

I owe a debt of gratitude to many people without whose help and guidance this book would not have been possible: my two fathers who taught me to pursue truth even when it runs contrary to the opinions of your friends, neighbours, and those in positions of power and authority; my mother who taught me that I could achieve anything I set my mind to, my wife for showing me that one can be a believer without being consumed by religion, my children for being the best kids in the world, my master's thesis advisor at the University of Saskatchewan, Eric Neufeld, who taught me how to write scientific papers; my social science colleagues who were kind enough to listen to my half-baked ideas and patiently explain social science concepts in terms that even a computer scientist could understand! - including members of the short-lived I-75 Cognition and Culture Research Group (Ryan Tweney, D. Jason Slone, Lauren Gonce, Kristin Edwards, and Joel Mort), Michael F. Young, Rik Warren, Mary Harmon-Vukic, and Lisa Legault. My interactions with them (especially with Ryan Tweney) provided exactly the support I needed to make the transition from computer scientist to cognitive scientist.

I am also thankful to all those who took the time to read through early versions of this book (which were much harder to read than the version in front of you, thanks to their feedback): Mansoor Ahmad, Amtul Rauf, Bushra Upal, Hinna Upal, Justin Lane, two anonymous reviewers, and the language editor at DeGruyter. My lifelong friend Mansoor was particularly helpful in obtaining access to a number of original sources. I am also thankful to my editor Rasa Pranskevičiūtė who conducted two rounds of thorough reviews that resulted in great improvement in the quality of the book.

1 Introduction

I grew up calling two men my father. In the West, where many people have a biological father and a step-father, this is not as unusual as it was in Pakistan where I grew up. I did not meet my biological father, Moulana Ali Haidar Upal Shaheed[1], until I was seven years old. The only contact with my father during my early childhood was his weekly letters and the occasional black and white photograph that he mailed us from West Africa where he was working as a missionary. My mother says that my primary motivation for learning to read was to be able to read *Abbijan's* (Urdu for father) letters on my own instead of having to beg her to read them to me. The older man in whose house we lived and whom I called *Abbaji* (also meaning father in Urdu) was actually my maternal grandfather. I called him Abbaji because my mother called him Abbaji. I also called my maternal grandmother *Ammiji* (Urdu for mother) following my mother (whom I called *Ammijan*, also meaning mother). Rather than correcting me, my grandparents took pleasure in my childish behavior perhaps because they didn't have a son of their own. In the fiercely patriarchal Pakistani culture not having a son is almost as bad as being childless.

Abbaji, my grandfather, had been a proud man in his youth. He was the first one in his village to go to college and the first one in his family to learn English. He graduated as an overseer from an engineering college and worked as a public servant first for the government of British India and then Pakistan. An estimated twelve million Hindus and Muslims died during the ethnic cleansing that followed the partition of the Indian subcontinent into India and Pakistan in 1947. My grandfather who was born and bred in East Punjab followed most Eastern Punjabi Muslims in abandoning his ancestral home for Western Punjab which became part of Pakistan. He avoided the fate of six million Indian Muslims, who were killed by Hindu and Sikh mobs during their ill-fated journeys to Pakistan, in part because he was a member of the well-organized *Ahmadiyya Muslim Community*[2]. Caravans of Ahmadi trucks moved together, sometimes under police protection, carrying thousands of Ahmadis from their headquarters in the East Punjabi village of Qadian to the West Punjabi city of Lahore, which became part of Pakistan.

Once in Pakistan, the Indian refugees were offered a chance to claim properties in Pakistan abandoned by Hindus and Sikhs in lieu of the properties they had left behind in India. The head of the Ahmadiyya Community, known as the *Khalifa* (Arabic for Caliph[3])-Allah's infallible viceroy on earth, advised Ahmadis against abandoning their Indian properties because Allah would soon victoriously return Ahmadis to Qadian-the small village where their faith had been born a bare half-century earlier in 1889. This meant that Abbaji and thousands of other pious Ahmadis started their lives from scratch in the new country. Abbaji did well enough to buy a house in the city of Gojra where he raised his three daughters. When the Khalifa (whom Ahmadis affectionately call *Hazoor*, meaning honourable) founded the new Ahmadiyya headquarters on the barren western shores of Chenab river in the early 1950s, Abbaji bought a tract of land there as his future retirement home. The modest home that he built in Rabwah came in handy when he decided to arrange for his middle daughter to marry an Ahmadiyya missionary school (called *Jamia Ahmadiyya*) student in 1967 who couldn't afford a home of his own.

Abbijan, my biological father, stood out among the Jamia students for a number of reasons, not the least of which was his age. In his mid-twenties, he was almost a decade older than some of his youngest class fellows. In order to become a Jamia student, one had to take a vow to offer all of one's capacities in the service of Hazoor without ever asking for anything in return. For a typical Jamia student, this was often done by the parents writing to Hazoor to seek his permission to offer their children to God. These students would then enter Jamia as soon as they finished their middle school. In order to induce their sons to keep their word to Hazoor, the parents would often promise to pay them a stipend for the rest of their lives. My paternal grandparents, who farmed in a small village in Sialkot, Pakistan, did not offer any of their six sons to the movement. Instead, they supported their second eldest son's decision to become the first person in the village to go to college and study sciences. After graduating with a degree in Physics and Mathematics, Abbijan took up a government job as a high school science teacher. It was after a few years of teaching that he received his calling from God and offered himself to the head of the community.

Because Abbijan had essentially neither any income of his own as a Jamia student nor a parental stipend, my mother had to work outside the house to support the family, an indignity for a middle-class Pakistani woman. This meant that as the movement sent my father from one Pakistani town to another following his graduation in 1970, my mother had to stay behind at her parent's house and close to her job. Thus, when I was born, it was Abbaji who wrote the letter to Hazoor (customarily written by father of the newborn) requesting him to bless the newborn with a name. Hazoor, perhaps thinking that the author of the letter (i.e., Abbaji named, Muhammad Sharif) was requesting a name for his own son, gave to me my grandfather's (and Prophet Muhammad's) first name and wrote with his own hands, "Muhammad Afzal, congratulations!" on Abbaji's postcard. More than forty six years have passed but that postcard is one of my mother's most prized possessions, not just because it is the first time my name was written, but also because it once touched the hands of Hazoor. "Afzal" can be either a comparative or superlative term in Arabic meaning better or best (depending on the context). Just as no English speaking parents would name their child Better/Best, Arabic speakers do not name their children Afzal!

By the time I was born, the persistent efforts of Ahmadi pioneers had managed to turn the inhospitable saline and rocky soils of Rabwah into the first planned city of Pakistan fulfilling the divine dream of Rabwah's founder, the second Ahmadi Khalifa (son of the founder of the Ahmadiyya Community who reigned from 1914 to 1966). The city of 40,000 entirely populated by Ahmadis served as the headquarters of the millions-strong worldwide Ahmadiyya Muslim Community. Unlike Qadian where Ahmadis had to cohabit with a large number of Hindus and Sikhs, Rabwah was a city built from scratch by Ahmadis for Ahmadis. The Ahmadi officials, all personally appointed by Hazoor, not only served as spiritual guides but also as the city's civil administration. Rabwah was divided into about thirty precincts with a mosque serving as the focal point for each precinct. The community enforced complete gender segregation with separate schools and colleges for men and women. Women had to cover themselves from head to toe in a black *burka* with only their eyes showing when they had to leave their houses. The city had no cinemas because even Pakistani films promoted immorality by showing unveiled women. Anyone caught visiting cinemas in the neighbouring city of Chiniot (or caught violating any of the other numerous bans such as the ban against flying kites, the ban against males clapping publicly, the ban against celebrating birthdays, the ban against associating with non-Ahmadis e.g., by taking part in their weddings or funerals[4] etc.) was punished by excommunication and/or expulsion from Rabwah. To encourage people to attend the daily Islamic ritual prayer of *Fajr* (offered at pre-dawn) groups of boys and young men went around shouting "prayer is better than sleep" at pre-dawn hours in the streets of Rabwah. We, the residents of Rabwah, took great pride in our status as the most organized and the most disciplined of all Muslims who were ready to sacrifice our lives to ensure a fulfilment of God's eschatological plans of making Islam the dominant religion before the end of the world. At our weekly boys' association meetings (and later at men's association meetings, as I grew older), we eagerly chanted our pledge "to sacrifice my life, wealth, time and honour for the sake of our faith" and "for guarding the institution of Ahmadiyya Caliphate."

While the community sent my father to various parts of Pakistan and to other countries, we lived with my mother and maternal grandparents in Rabwah, where my mother taught at a girl's primary school. My father's first missionary appointments were in various Pakistani cities so he must have seen me on one of his visits home but I have no memory of that. When I became old enough to remember faces, he was working in the West African nation of Ghana (where Jama'at had sent him in 1973). I learned of my father's face from the few black and white photographs that he had mailed us from Africa. These pictures looked eerily similar to the author's picture on the back cover of the well-known Ahmadi Urdu book on the shroud of Turin. In his book, Hasan Muhammad Khan Sahib, a local celebrity, defended the heretical Ahmadiyya doctrinal position on Jesus' crucifixion. Imagine my surprise when I finally saw my father for the first time, at the age of 7, and he didn't look anything like Hasan Muhammad Khan Sahib! My father stayed with us in Rabwah for about six months before being sent to Mirpur Khas in Sindh province of Pakistan. The few times that I got to see my father between his various missionary assignments, I never heard him complain about having to sacrifice his family life. In fact, I heard countless times from him and from other Ahmadis that sacrifices for the community were the real cause of any success, however small, that our family had had. My good health, my good grades in school, and my success in constantly ongoing religious competitions in Rabwah were all a direct result of my father's missionary vows[5].

The early 1970s were a heady time for Ahmadis in Pakistan. Ahmadis believed that they had played a decisive role in the victory of the socialist Pakistan People's Party (PPP) in Pakistan's first elections based on universal suffrage held in December 1970[6]. They believed that PPP would pay them back for their support by allowing them to preach unmolested which, in their minds, could only lead to one outcome, namely, a quick acceptance of *Ahmadiyyat* (Arabic for Ahmadism) as true Islam first by all Pakistanis and then by the rest of the world. This elation turned into a deep sense of betrayal in 1974 when Prime Minister ZAB Bhutto capitulated to the Sunni mullahs demanding a constitutional amendment to declare Ahmadis non-Muslims. When Bhutto was hanged by the military in 1979, the atmosphere at the Annual Ahmadiyya Convention, that brought hundreds of thousands of Ahmadis from across the globe into Rabwah every year for a long weekend in December, was openly celebratory. There was a strong sense that prophecies predicting destruction of any worldly power that took on God's chosen people had been fulfilled in front of our eyes.

With the centennial of the community's founding only a decade away, we were eagerly anticipating people joining Ahmadiyyat in droves. I remember frequently running into the community's official historian Moulana Dost Muhammad Shahid (who worked at Khilafat Library, two blocks from our new house in the Foreign Missionaries' Colony where we moved in 1983[7]). Every time I saw him, he would ask me what would my age be in the year 2000 when people around the world would be flocking to Ahmadiyyat. Will I be ready to accept the hordes of non-Ahmadis desperately looking for anyone knowledgeable enough about Ahmadiyyat to save their souls? The fact that the world was going to be saved through Ahmadiyyat was not in any doubt. What was doubtful was whether we, as individuals, were willing to play our role (and gain salvation for ourselves). At the daily religious lectures that were delivered following the evening prayers at our neighbourhood mosque, at our weekly *Atfal* (Young Ahmadi Boys' Association) meetings, at the yearly Quranic-education classes, and at the Annual Ahmadiyya Conventions, we were routinely told about the success that Ahmadi missionaries, such as Abbijan, were having in converting non-Ahmadis to Ahmadiyyat around the world. This success proved that God was on our side.

While both my "dads" were proudly faithful Ahmadis, their faith-styles differed from each other. Abbijan had the cerebral faith of a born-again scholar, having heard God's calling after studying and teaching science for years. My father had spent seven years in Jamia Ahmadiyya Missionary School learning the minutia of theological arguments. He had travelled around the world engaging people of all faiths in religious debates. This set him apart from most other Ahmadis (such as my Abbaji) who hadn't read all the books of Hadith or Ahmadiyya doctrine or traveled widely but had a strong emotional attachment to the Jama'at and a firm belief in the supernatural. This included a belief in the miraculous powers of the Khalifa and an irrational reverence for anyone (such as the family of the founder of the Ahmadiyya Jama'at) or anything connected to the founder of the Ahmadiyya Jama'at. My grandfather would get extremely upset if he heard anyone say anything that he perceived as insulting towards a member of the *khanidan* (hundreds-strong progeny of the founder of the Ahmadiyya movement who lived in a gated area in central Rabwah).

Similar to most devoted Ahmadis, my grandfather endeavored to give ever more of his meagre resources to the community. Not content with contributing the obligatory 1/16th of his modest government pension to the movement, Abbaji wrote to Hazoor to grant him permission to donate 1/8th. Once he was allowed to do that, he wrote another letter asking for permission to contribute 1/4th of his income. Once that request was granted, he asked the proportion to be increased to 1/3rd. Hazoor, to his credit, denied that request. Abbaji refused to take 'no' for an answer and kept writing (to be more accurate, he kept asking me to write because by this time he was too old the write) until he had his way. Had he lived longer, I'm sure he would've requested permission to donate ½ of his income and eventually all of it! In 1979, when the third Ahmadi Khalifa (grandson of the founder of the movement who reigned from 1966 to 1982) launched the decade-long campaign to raise funds to celebrate the centennial of the founding the Ahmadiyya movement in 1889, Abbaji sold his house in Gojra and donated the entire proceeds to the Ahmadiyya Centennial Fund. I don't think I ever saw him happier than he was on the day he received an invitation to a private meeting of major donors with Hazoor. He had my mom put starch on his new *shalwar kameez* (long shirt and tunic) and his white turban that he proudly wore to the meeting with Hazoor. His face beaming with happiness at the prospect of being able to spend time in the physical presence of God's infallible regent on earth.

While I barely knew my absent father, my grandfather was my role model. Like him, I would get up well before sunrise to offer the optional *tahajjud* prayer. I prayed for Allah to make me a success both in the secular and the spiritual world, like my Abbaji. I wanted to be ready to play my part in God's plans for the end times. During the Jalsa when we had a house full of relatives from across Pakistan, I led everyone in the ritual prayers, just as Abbaji often did at our neighbourhood mosque. I read the Quran (first in Arabic and then in translation in Urdu), first from Abbaji, and then from Moulana Zakriya Khan Sahib–the community's official Albanian language translator. I read the books by the founder of the community and his successors (i.e., the Khalifas). I eagerly enrolled in the annual *Talim-ul-Quran* (Quranic-education) classes and took part in competitions of religious knowledge, rhetoric, and Arabic recitation. Above all, as did my Abbaji, I internalized being a good believer as a critical part of my identity. I still remember some of the verses of my favourite childhood poem that I used to sing to anyone who would listen:

hum Ahmadi bache hain kuch kar ke dikha dain gay [we're Ahmadi kids, we will do something big]

shaitan ki hakumat ko dunia se mita dain gay [we will abolish Satan's government from the earth]

aey Shad guman mat kar, kamzor nahin hain hum [don't worry for us, we are not weak]

jab waqt para hum apni janain bhi ganwa dain gay [when the time comes we'll even sacrifice our lives] hum Ahmadi bache hain kuch kar ke dikha dain gay [we are Ahmadi kids, we will do something big]

Islam ko dunia main mehboob bana dain gay [we will make Islam everyone's beloved religion]

I think I was seven when I memorized the whole poem and sang it for a delighted Abbaji, who kissed me on the forehead.He had the face of a revolutionary filled with the satisfaction of having passed his revolution to his "son."

I always sided with my grandfather when, on occasion, his strict interpretations of Islam, contradicted with, what seemed to me back then, relatively more liberal interpretations of my parents. My mother enjoyed listening to Indian and Pakistani songs on radio when she cleaned the house. This used to irk Abbaji who often found the lyrics to be lewd and distasteful. According to him, the only acceptable use of a radio was to listen to news (Radio Pakistan news and BBC Urdu service news that came at the turn of every hour). As his eyesight became weaker, I became his eyes and hands. I mastered the fine art of tuning the radio to receive the hard-to-catch BBC Urdu signal. I also decided to do a bit of research at the Jama'at library and found *ahadith* (sayings of the prophet Muhammad) expressing displeasure with music. With Abbaji's approval, I plastered handwritten notes with these ahadith all over our house in Rabwah. I did the same thing at the missionary's quarters in Karachi in 1979 posting anti-smoking ahadith everywhere when I heard that my father enjoyed an occasional cigarette when he was alone (he never smoked in front of us). On such occasions, my parents made fun of my overly dogmatic attachment to religion by calling me a "moulavi" (i.e., an Islamic cleric) and "imam-manja" (i.e., the prayer leader of the beds"). At other times, when I was not annoying them with my self-righteousness, they asked me to pray for them because they said God especially listened to my prayers. Somehow I had acquired the identity of being the most devoted Ahmadi in a family of extremely devoted Ahmadis.

Ahmadis believe that "belief in a living God" who talks back when people call on him distinguishes them from non-Ahmadi Muslims who believe that God has only communicated with holy personages in the past. Cultivating an active communicative relationship with God is an aspirational goal for all Ahmadis. From my grandfather, I also inherited the sense of a personal relationship with God. Throughout my childhood I felt that not only was God listening to my prayers but that he was also actively telling me things by showing me his signs. One of the signs of God's special favours for me was shown to my eighth grade public school teacher, Master Nazeer Sahib. Just a month or so before our dreaded Middle School Standard Exams, our homeroom teacher at the Talim-ul-Islam School, called me to the front of the class to tell the whole class about a supernatural experience that he had had in the early hours of that morning in February 1983. He said that as he was walking in the dark from his house to the mosque for the morning prayers, he heard a distinct voice saying "Muhammad Afzal, scholarship." This, he said, was Allah himself foretelling him that not only would I top my competitors in Section A but that I was destined to score high enough marks to earn the coveted government scholarship in high school. The fact that the prophecy revealed to Master Nazeer Sahib came true and I won top positions in middle school, and then high school and college, as well as in the religious competitions organized by the community[8] only reinforced my identity as God's chosen one. Growing up poor on the mean streets of Rabwah without an older brother or a protective father, I found security in the belief that I could always count on Allah. It was comforting to know that there was someone there who always had my back. I wish that I could describe in words the sense of purpose, meaning, and joy that comes from knowing that you have a direct channel of communication with the creator and all-powerful lord of the universe.

Sometimes, late at night, especially when I'm having trouble going to sleep, I think about how I lost that feeling and I cannot point to any single event but I think that it all started with the typical teenage rebellion and Abbaji's passing away in 1989 (the Ahmadiyya centenary year that he had so longed to see). My grandfather's death was one of many factors (including an opportunity for my mother to take early retirement with government pension, a decline of Rabwah's status as a mecca for Ahmadis following Hazoor's move from Rabwah to UK in 1984, and general economic malaise in Pakistan which made moving to the West a trend among Pakistani Ahmadis) that convinced my parents that we should join our father in Canada. As luck would have it, by the time, we got our immigration papers and arrived in Saskatoon, Canada, Jama'at had decided to send Abbijan back to Africa (this time to the small West African country of Gambia). This meant that as I arrived in Canada I had to find my own way in a radically different world.

Saskatoon of the early 1990s seemed as different a place from Rabwah as possible. When we arrived there in December 1990, streets were covered with some white powder (locals called it "snow") instead of the brown dirt that littered Rabwah's streets. When the sun came out, the weather got colder and not warmer. People drove on the right side of the road and not the left side. People called football "soccer" and they called something resembling group-wrestling football. Women had shorter hair and wore more revealing clothes than men. Everything seemed upside down. The most bewildering thing was that this opposite-world somehow seemed to work better. My father put it best when he said that if the hadith "cleanliness is half the faith" is to be believed then clearly Canadians are better Muslims than Pakistanis! Many of the egalitarian ideals that Ahmadis claimed to aspire to, such as justice and equality of the rich and poor, seemed closer to being met in the West than they ever were in Rabwah. Instead of finding any Western converts (that I expected to find given what we were told in Rabwah about the success of Ahmadi missionary efforts in the West), the only Canadian Ahmadis, I met were Pakistani Ahmadis who had immigrated to Canada. Instead of finding droves of non-Ahmadis wanting to convert to Ahmadiyyat, I found Canadians, by and large, apathetic to the message of Ahmadiyyat and Islam. Instead of finding immorality, injustice, and moral filth in the West (as proclaimed by Ahmadi leaders as well as by many other Pakistanis) I found the West to be a pretty nice place to live. All of that shook up my worldview and made me question that maybe everything I was told was not as true as I thought that it was. This made me open to the West's liberal narrative that the road to egalitarianism rests on a foundation of constantly questioning authority and not in blindly following it.

When I enrolled in the University of Saskatchewan's computer science program, I was required to take social science courses to fulfil my breadth requirements. I chose Philosophy and Anthropology classes. I found the social science and humanities approach of adopting multiple perspectives quite refreshing. This was opposite to the experience of many of my fellow computer science students who were baffled by the lack of consensus among social scientists on issues of fundamental importance to the disciplines. Afflicted by typical immigrant fears of not finding a job, I just could not muster the courage needed to change my major to social sciences. After completing my PhD in Computer Science, however, I continued my own reading of Psychology, Anthropology, and Philosophy while working as a computer scientist. The more I read about the theory of evolution, the less certain I became of the traditional Islamic explanations for the beginning of life and human existence.

Most of the Ahmadis who move to the West do not go through this transition and I have often wondered why not? Why do people continue to hold blind faith in cult leaders in an era of instant information and social media? I have spent most of my adult life contemplating these questions as well as other questions that my friends and family ask me. Not a day passes by that my mother, my in-laws, my siblings, or my friends don't ask me as to why I am not as active in the Ahmadiyya community as I used to be during my childhood. Why don't I take a place of honour, distinction, and leadership in the community that my talents and lifetime of achievements would surely bring me? My family and friends genuinely do not understand how I could have so easily given up the blessings of Ahmadiyyat and lost my enthusiasm for "true Islam." My scientific colleagues, on the other hand, do not understand how and why any rational person could believe in cultish ideas such as those proclaimed by the founder of the Ahmadiyya Community and held in such high esteem by most of my family and friends. Only simpletons can be brainwashed by cult leaders into blindly following them, argue my scientific colleagues. This book is both a deeply personal account of my struggle to find answers to these puzzling questions and a scientific account of why new religious movement founders come up with their counterintuitive ideas and why others accept them and join new religious movements.

1.1 Ahmadiyya Muslim Community

The founder of Ahmadiyya Muslim Community, Mirza Ghulam Ahmad, was born in the Punjabi village of Qadian around 1839. He was the younger son of the village chief, Mirza Ghulam Murtaza. His family was part of the Central Asian armed bands that moved eastward to India along with Mughals who ruled India for centuries before losing most of their territory to Marathas and Sikhs in the eighteenth century.

Ahmad's family lost their estate (including their ancestral home in their capital of Qadian) to the Sikhs, only to be allowed back around the time of Ahmad's birth. By the time Ahmad was ten, the Sikh rule over Punjab ended marking the culmination of the Anglo-Sikh wars of 1846-49. After the British crushed the 1857 mutiny of Indian soldiers, a quick restoration of Muslim dominance over India seemed impossible. The British occupation also hastened the pace at which Punjab was exposed to new ideas and technologies. These included schools and colleges that taught an ever-larger number of Indians how to read and write and new technologies such as the printing press that allowed mass production of books, pamphlets, and journals in English as well as local languages, particularly Urdu.

Ahmad, as the younger son of a feudal family, was well positioned to become an early adopter of the new media because he had the luxury of time and money. Ahmad, while in his thirties, started writing articles defending Islam against objections by Hindu revivalist movements and Christian missionaries. These efforts were lauded by the small literate Indian Muslim elite. This encouraged Ahmad to dream big and launch the ambitious project of publishing a 50-volume book series that would comprehensively rebut all Hindu and Christian arguments against Islam. The first volume of the series, *Braheen-e-Ahmadiyya*, published in 1884, stated that the goal of the series was to establish Islam's superiority over all other religions (particularly Christianity and Hinduism). One of Ahmad's key arguments was that, unlike Christianity and Hinduism, Islam endowed its faithful followers with the ability to communicate with a living God who supported them through his signs. Ahmad offered evidence of this in his own dreams that had come true and that divine revelations he had received about many future events, including the death of his father. Ahmad challenged Hindus and Christians to come to Qadian and live there at his expense for at least a month so that they could witness firsthand how God still showed his signs through him. While some Muslim clerics were bothered that Ahmad was claiming too much spiritual powers (especially the power to show miracles) for himself, others lauded Ahmad's noble attempts to defend Islam at his own expense. Either way, Ahmad was unapologetic in insisting that he was doing all this only to prove Islam's superiority and to save Muslims from converting to Hinduism and Christianity.

In the late 1880s Ahmad came to believe that Christianity was the bigger threat facing Islam and that traditional Islamic doctrine had left Islam vulnerable to Christian attacks. He believed that these doctrines needed to be reformed. Ahmad blamed the Muslim belief in Jesus having escaped crucifixion by being raised to heaven as the key culprit. He said that Evangelical missionaries were arguing that this showed that Jesus was superior to Muhammad, who lay buried underground in Medina. Missionaries reminded Muslims of their own eschatological beliefs in Jesus' physical descent from heaven towards the end of times to bolster their arguments for the superiority of Jesus over Muhammad. Ahmad claimed that God had told him that Jesus had survived the indignity of crucifixion and had traveled to India where he died a natural death. Furthermore, he said that since the old Jesus was dead and could not return, God had appointed him as the Messiah of the end times because he was similar to Jesus in many ways. These claims lost Ahmad any remaining support among Muslim notables who now almost uniformly criticized him as an innovator and a false prophet. Ahmad felt the need to call on his supporters to sign a formal oath of allegiance to him. Such oaths had traditionally been used by Sufi teachers to formally accept those seeking to learn from them as their students. Forty people gathered in the Punjabi city of Ludhiana on the morning of March 23, 1889 at the house of one of Ahmad's supporters to sign their pledges. A decade later he asked his followers to register themselves as *Ahmadiyya Muslim Jama'at*[9] in the 1901 India census, cementing their status as a distinct religious community.

Ahmad made paying 1/16th of one's income as a condition of membership in the Jama'at and established organizational offices (called *sadr-anjuman*) in Qadian to administer the funds. He established a press in Qadian that published Urdu and English periodicals and books. He also established the tradition of holding annual Ahmadiyya conventions (called *Jalsa Salana*) in Qadian in December. Despite almost universal opposition by Muslim clerics, or perhaps because of it, Ahmad continued to gather new converts. He claimed thousands of middle class followers by the time of his death in 1908.

Ahmad's best friend, and his biggest financial benefactor, Hakeem Nur-ud-Din, assumed Jama'at's leadership following Ahmad's death as the first Ahmadi Khalifa. After Nur-ud-Din's death in 1914, the community split up into two factions. While Ahmad's son Bashir-ud-Din Mahmud Ahmad assumed the role of the second Khalifa of the larger faction of Ahmadis who stayed in Qadian, a smaller number of Ahmadis moved to Lahore and set up a rival faction. The Ahmadiyya Muslim Movement of Lahore claimed that Ahmad was not a prophet but a lower-ranked reformer (called *mujaddid*) and hence he should not be succeeded by a Khalifa but by an administrator. When the British partitioned their South Asian colony in 1947 into Pakistan and India, Qadian fell on the Indian side of the border. This forced the Qadian faction to move their headquarters to Pakistan (first to Lahore and then to Rabwah). After Bashir-ud-Din's death in 1966, his eldest son Mirza Nasir Ahmad assumed the role of third Khalifa. On September 7, 1974, Pakistan's parliament unanimously adopted the second amendment to the constitution to declare Ahmadis non-Muslims. On Nasir Ahmad's death in 1982, his step-brother Mirza Tahir Ahmad became the fourth Khalifa. On April 26, 1984, Pakistan's military dictator issued an ordinance (later approved by Pakistan's parliament) which made Ahmadiyya proselytization (an integral part of the faith) a crime punishable by a jail term and a fine. Three days later, Tahir Ahmad moved to London to avoid prosecution under the new law. In 1989, he dissolved all international Ahmadi organizations formerly headquartered in Rabwah and moved many Jama'at offices to Tillford, UK[10]. Upon his death in 2003, Mirza Ghulam Ahmad's great grandson, Mirza Masroor Ahmad became the fifth Khalifa. He resides in London, UK.

Table 1 shows key differences between Ahmadiyya doctrine and mainstream Sunni Muslim doctrine. Table 2 show a brief timeline of key events in Ahmadiyya history.

Table 1: A summary of key doctrinal differences between mainstream Sunni Muslims and Ahmadi Muslims

Issue	Mainstream Sunni Muslim Position	Ahmadiyya Position
Who is Allah's final prophet on earth?	Muhammad was the final prophet of God. There can be no prophet after him.	There are two different types of prophets: those who bring a new religious system (i.e., a *shariah*) and those who come to restore an old one. While Muhammad was the final law-bearing prophet, Ahmad was only a reformer prophet who came to restore the true Islam.
Was Jesus (whom all Muslims consider to be a prophet) raised to heaven alive?	Yes. God saved Jesus from the indignity of crucifixion by raising him to heaven while he was still alive and by making someone else look like him. The lookalike was hanged on the cross.	No. Jesus was the one who was hanged on the cross but he only suffered minor injuries. He recovered and traveled to India to continue his mission to the lost tribes of Israel who lived in Afghanistan and Kashmir. He died at the age of 120 and is buried in Srinagar, India.
When will Jesus return to earth?	Jesus is alive in heaven. He will descend from the heaven towards the end of times to fight the anti-Christ and establish Islam as the dominant religion worldwide.	Jesus is dead & will never return to earth. The Messiah whose return in prophesized in *Ahadith* is someone who is similar in character to Jesus. The Jesus-like Muslim Messiah will fight the anti-Christ and establish Islam as the dominant religion.
Who is the Imam Mahdi?	Jesus and Imam Mahdi will be two different people. Jesus, similar to other Muslims, will follow Imam Mahdi as his guide. Neither leader has appeared yet.	Imam Mahdi and Jesus were supposed to be the same person. Both these prophecies were fulfilled in the person of Mirza Ghulam Ahmad- the founder of the Ahmadiyya Jama'at.

Table 2: A timeline of key events in Ahmadiyya history

Date	Event
1831	Syed Ahmed Barelvi, whom some Indian Muslims considered Mahdi is slain by the Sikhs during a battle at Balakot
1839[11]	Mirza Ghulam Ahmad (MGA), founder of Ahmadiyya Muslim Jama'at is born
1846-49	Anglo-Sikh wars won by the British result in the British takeover of Punjab
1854	MGA is married by his parents to his cousin from whom he has two sons
1857	The mutiny of Indian soldiers claiming to restore the last Mughal King in Delhi is crushed and the King is exiled to Burma. Queen Victoria assumes direct control of India
1876	MGA's father, Mirza Ghulam Murtaza, dies in Qadian
1879	MGA announces the launch of his "Braheen-e-Ahmadiyya" book series

Continued Table 2: A timeline of key events in Ahmadiyya history

Date	Event
1884	Publication of first two volumes of "Braheen-e-Ahmadiyya" & MGA's second marriage to the daughter of a prominent *Ahl-e-Hadith* (a reformist Islamic sect) family from Delhi is arranged by his friend & *Ahl-e-Hadith* leader Moulavi Muhammad Hussain Batalavi
1889	MGA has forty of his followers sign pledge of allegiance laying foundations of the Ahmadiyya Muslim Jama'at (AMJ)
1891	MGA publishes three books ("Fath-e-Islam," "Tauziay Maram," & "Azala-Auoham") proclaiming that Jesus survived crucifixion and died of natural causes in his old age. He also claimed to be both the prophesized Mahdi and Messiah. MGA announces the Annual Ahmadiyya Convention (*Jalsa Salana*) to take place in the last week of December in Qadian
1895	An Ahmadiyya Press is established in Qadian
1897	An Urdu Ahmadiyya periodical "Al-hakm" and a boys' school (called Talim-ul-Islam School) is established in Qadian
1899	MGA publishes "Jesus in India" outlining his theory that Jesus traveled to India after recovering from his injuries on the cross, died at the age of 120, and is buried in Srinagar, India
1901	MGA asks his followers to register themselves as "Ahmadiyya Muslim Jama'at" in the 1901 British India census
1902	The English language periodical "Review of Religions" is launched to preach Islam to the West
1905	MGA publishes one of his last books titled "The will" creating a special category of tithe-paying Ahmadis who will be buried in the *Heavenly Cemetery* initially established on the grounds of his family cemetery in Qadian
1906	MGA establishes Jama'at's administrative offices called *Sadar Anjuman Ahmadiyya* in Qadian
1907	MGA asks Ahmadis to take missionary vows to dedicate their lives to propagation of Islam in India and around the world without asking for any compensation in return
1908	MGA passes away and Maulavi Nur-ud-Din assumes the title of Ahmadi Khalifa
1912	The first Ahmadiyya missionary to the West, Khawaja Kamal-ud-Din, establishes a mission at Woking Mosque in London
1914	Maulavi Nur-ud-Din passes away and the movement splits into Lahori and Qadiani sects[12]. Ahmad's eldest son from his second wife, Mirza Basheer-ud-Din Mahmud Ahmad, assumes the title of 2nd Khalifa in Qadian while the Lahori Ahmadis abolish Caliphate
1947	Qadiani Ahmadis move their offices to Pakistan (temporarily to Lahore and eventually to Rabwah)
1948	Mirza Bahir-ud-Din identifies the barren rocky land on the western shores of Chenab river as the divinely promised land for a new headquarters shown to him in a revelation and names it Rabwah

Continued **Table 2:** A timeline of key events in Ahmadiyya history

Date	Event
1953	Anti-Ahmadiyya riots break out in Pakistan's Punjab province (where Rabwah is situated and most Pakistani Ahmadis live) leading to declaration of martial-law in Punjab and military's crushing of rioters
1965	Mirza Baheer-ud-din Mahmud Ahmad passes away. His eldest son Mirza Nasir Ahmad assumes the role of 3rd Khalifa
1974	Anti-Ahmadiyya riots break out in Pakistan again and end only after parliament passes a constitutional amendment declaring Ahmadis to be non-Muslims
1982	Mirza Nasir Ahmad passes away. His step-brother Mirza Tahir Ahmad assumes the role of the 4th Khalifa
1984	Pakistan's Martial-law administrator, General Zia-ul-Haq, issues an ordinance making it illegal for Ahmadis to engage in Islamic practices and preaching (passed into law by Parliament in 1986). Mirza Tahir Ahmad moves to England to establish movement's new offices in Tillford, UK
2003	Mirza Tahir Ahmad passes away. His nephew, and MGA's great grandson, Mirza Masroor Ahmad assumes the role of the 5th Khalifa. He leads a millions-strong[13] worldwide Ahmadiyya Muslims Jama'at from London, UK.

1.2 Scholarly Studies of the Ahmadiyya Muslim Jama'at

The origin and spread of Ahmadiyya Muslim Jama'at has confounded both historians and scholars of religion. "A curious religious phenomenon in Indian Islam has been the advent of Ahmadiyya movement," wrote the eminent scholar of Islam Fazl-ur-Rahman. "Historians remain baffled as to how a group that is regarded as so outside mainstream Islam attracted enough followers to remain viable until now", said scholar of religion Shazia Ahmad (2010). Reviewing Friedman's book "Prophecy Continuous" (which deals with Mirza Ghulam Ahmad's various claims), Fusfeld (1992) wrote:

> The relationship between the Ahmadiyya movement and the political, economic, and social environment (as distinct from its intellectual origins) is . . . largely unexplored. There is never any satisfactory explanation offered to show why the founder of the Ahmadiyya movement chose to take positions that were so outlandish when viewed from the perspective of mainstream Islam. . . how he benefited from taking a position on the finality of prophethood that many other Islamic leaders viewed as beyond the acceptable boundaries of Islam. . . why the Ahmadiyya came to be in such an unorthodox and (from the point of view of other Muslims) unacceptable positions... (Fusfeld, 1992: 347-348)

Fusfeld faulted the traditional approach to new religious movements for its shortcomings to explain their origin and growth by exclusively appealing to sociological factors. "The Ahmadiyya movement was to a large extent the result of one person's view of the world he found and his efforts to come to grips with the problems he perceived. If the solution was a peculiar one, it may owe its peculiarity to the person who made it work" said Fusfeld (1992: 348).

Ahmadiyya is not the only religious movement to have been influenced so profoundly by the radically innovative ideas of its founder. Many new religious movement (NRM) founders (such as Joseph Smith, Sun Myung Moon, and Mary Baker Eddy) appear to have based their respective movements on similar innovations. Joseph Smith (1805-1844) claimed to be "a prophet and apostle of Jesus Christ". He also claimed to have received divine knowledge of the journey of a group of Israelites into the Western hemisphere. Smith believed that he was given special powers to interpret this lost knowledge into English. Using these powers, he composed the *Book of Mormon* which is considered scripture by most members of the Church of Latter Day Saints. Mary Baker Eddy (1821-1910) claimed to be a "God-appointed messenger" who was chosen by God to give "full and final revelation of truth." According to the doctrine of the Church of Christian Science, "she is so closely related to Christian Science that a true sense of her is essential to the understanding of Christian Science; in other words, the revelator cannot be separated from the revelation". Similar to Mirza Ghulam Ahmad, Sun Myung Moon (1920-2012) also claimed to be the second coming of Jesus. He claimed titles of "the Messiah, the Lord of the Second Advent," and "father of the universe". Rev. Moon also claimed that Koreans were descendants of the lost tribes of Israel (Hodge & Patterson, 2015).

It is clear that no explanation for the radically-innovative ideas of these new religious movement (NRM) founders can be complete without looking inside the heads of these individuals. A scientific explanation must be able to answer questions such as how and why do NRM founders invent radically new ideas, how and why do they communicate these ideas to others, and how and why others come to place their faith in these ideas. In a 2005 article "towards a cognitive science of new religious movements" I argued that that this can be accomplished by complementing the sociology of new religious movements by a cognitive science of new religious movements (Upal, 2005b). The first goal of this book is to fill more details into the broad outline of the theory presented in that article, by developing a multidisciplinary theoretical framework drawn from cognitive science of religion and social psychology (in particular social identity theory, and leadership as social identity change entrepreneurship). The second goal is to illustrate how this socio-cognitive account of the origin and evolution of NRMs can be used to understand origin and evolution of a real-world NRM, namely, the Ahmadiyya Muslim Jama'at.

1.3 Approaches for Studying New Religious Movements

Religion is a one of the most fascinating aspects of human life. Since time immemorial, it has provided answers to ultimate questions of existence for some people from a variety of cultural traditions. It motivates some people to offer sacrifices for the good of their fellow human beings. It provides a sense of meaning to many people's lives. It brings communities together by binding people in social ties of friendship. While questions of the origins of new religious ideas may perplex modern people living in largely secular societies, they have not been systematically investigated until the rise of modern social sciences. Studies of new religious movements gained strength in the West with the emergence of eastern-inspired movements in the 1960s. While there is no consensus on what constitutes a new religious movement, the label is usually reserved for religious movements of historically recent origin (founded in the last 200 years or so) that "have been assigned to the fringe of the dominant religious culture" (Lewis, 2008: 17). Bainbridge and Stark (1979) categorized the existing research on the origin of new religious ideas into three groups: the psychopathology model, the entrepreneur model, and the subculture evolution model. While the first two models emphasize the role played by individuals (particularly the NRM founders), the subculture evolution emphasizes the group interactions processes that can lead to the emergence of NRMs without guidance by any individual. I will spend some time on the first two models because they are relevant to the development of a bottom-up model of religious innovation.

According to the psychopathology model, NRM founders suffer from various mental illnesses which cause them to have psychotic episodes resulting in hallucinations of having received divine wisdom. Havelock Ellis suggested that most of the Israelite prophets were psychologically ill when he said that, "the whole religious complexion of the modern world is due to the absence, from Jerusalem, of a lunatic asylum," a view that is consistent with Freud who considered religion to be a projection of neurotic wish fulfillment which should be treated with therapy (Hockney, 2014; Freud, 1964). Religion Scholar George Feuerstein argues that NRM leaders are authoritarian personalities who want to control their followers (Feuerstein, 1991). After examining a number of religious movement leaders including Shabbatai Zwi, Jim Jones, David Koresh, and Rudolph Steiner, the British Philosopher Colin Wilson concludes that NRM leaders have psychological problems similar to those suffered by serial killers, namely, that they are driven by power and sex beyond the normal limits (Wilson, 2000). A number of psychological problems have been alleged to result in the claims of special divine insight by religious leaders. These include epilepsy (Saliba, 2004), hysteria (Zweig, 1932), paranoia (Gardner, 1957), and schizophrenia (Storr, 1996). Mary Baker Eddy's inspiration of Christian Science is often cited as the classic case of hysteria. The British Psychiatrist, Anthony Storr, argued that people such as Rudolf Steiner, Paul Brunton, Ignatius of Loyola, Gurdjieff, Rajneesh suffer from a "creative illness" that leads them to the creation of novel ideas.

When the American scholar, Dr. Hervey De Witt Griswold (I'll tell you more about him later), asked an Indian Muslim about Mirza Ghulam Ahmad's claims, he was told that:

> Mirza Sahib's brain has become meddled ("US KA DIMAGH BAITH GAYA"). This opinion concerning the Mirza Sahib is similar to the opinion of Festus concerning Paul... In connection with the theory of the Mirza Sahib's insanity, it may not be without interest to mention that at

> least two persons in the PANJAB, who are acknowledged to be insane, have lately claimed to be Jesus Christ, one a weaver of Ludhiana, and the other a former student of the Forman Christian College, Lahore. The madness of the latter takes the shape of writing periodical letters to the Principal of the College and urging that his claim to be the Messiah be speedily admitted. In the light of these facts, the theory that the Mirza Sahib himself is insane is certainly a possible one. (Griswold, 1902: 25)

Another unnamed American psychologist who also visited Ahmad was so convinced that Ahmad's revelations were a result of a brain disease that he directly asked Ahmad, "have you ever been affected with a brain disease? If so, what and when? Does its attack recur? Did you begin to have revelations before you suffered from an attack of such disease or after that?" After narrating this incident, Rev. Howard Arnold Walter opines that:

> I am indebted to Professor D. B. Macdonald, of Hartford, U.S.A., for the suggestion here advanced as perhaps best accounting for Ahmad's claims and so-called revelations, viewed in the light of our modern knowledge of psychology is that Ahmad like his great leader Muhammad, the founder of Islam, is a "pathological case"? It is evident that from comparatively early days he had trances; fell into fits in which he saw and heard strange things. There came to him voices, either apparently in a trance condition or when he was awake. Driven by fear for his soul, he had got into the habit of retiring into desert recesses, and there spending days in solitary prayer. So there the voices came to him; there he even saw figures—vague, dim—and the fear fell upon him, What are they? What is the matter with me? Is this of God? Or am I possessed by some spirit? (Walter, 1902: 45)

The image of religious innovators as social deviants is so well entrenched in popular culture, perhaps thanks to the tireless efforts of 'cult-hunters', that often little or no evidence is considered necessary to justify accusations of mental illness against them. However, as popular, widespread, and intuitive as this view is, we still must subject it to scientific scrutiny so that we do not fall prey to what philosopher Dan Dennett calls "premature satisfaction of curiosity" (Dennett, 2006: 103).

Let us take the mental illness hypothesis seriously and fully explore all its implications. One of the unfortunate consequences of mental illness is that the people suffering from the illness are stigmatized in a variety of ways. Surveys show that such people are wrongly believed to be stupid, lazy, unpredictable, unreliable and dangerous. This appears to be the case both in Western industrialized societies of North America and Europe as well as in non-industrialized societies of Asia and Africa. Unlike religious innovators whose adoring followers consider them as sources of divine knowledge, counterintuitive utterances of mentally ill people are ignored at best and ridiculed at worst. A recent survey of UK mental illness sufferers showed that a vast majority (70%) reported suffering stigma as a result of their illness. A majority also reported being stigmatized by their own family (56%) and friends (52%). Similarly, a vast majority of Americans (75%) consider mentally ill people to be 'dangerous' (Crawford & Brown, 2002). Furthermore, there is no historical or anthropological evidence that mentally ill people were respected for their wisdom and insight in the past. It is fitting then that life stories of religious innovators such as Ahmad show no history of a debilitating mental illness. Not only were they considered normal and healthy by their peers, they were actually respected by their followers (and sometimes even by their opponents) for their intellect, their wisdom and their insight. Dr. Griswold, for instance, describes that he found Ahmad to be "about sixty four years of age, venerable in appearance, magnetic in personality, and active in intellect" (Griswold, 1902: 1).

Newberg and D'Aquili (2008) in "Why God Won't Go Away" argue a subtly different position. They suggest that some founders of religions were 'mystics' who learned to experience a different state of consciousness through 'spiritual exercises.' This extraordinary state of cognition offered these people a unique view of reality which was 'more real than reality itself.' After studying brain scans of Tibetan Buddhists and Franciscan Nuns in meditation, Newberg and D'Aquili concluded that the meditative experiences of these mystics were not delusions or "the result of emotional mistakes or simple wishful thinking, but were associated instead with a series of observable neurological events." They conclude that "mind's machinery of transcendence may in fact be a window through which we can glimpse the ultimate realness of something that is truly divine" (Newberg & D'Aquili, 2008: 7).

Another common explanation for the behavior of religious innovators is that they are charlatans, frauds, and con artists who make sensational religious claims for financial and material gains. Religious scholars Shupe and Bromley (1981) summarize this view as arguing that religious movements are "profit-making ventures operated by egomaniac charlatans for their own personal aggrandizement" (Shupe & Bromley, 1981: 186). A problem with this view is that a vast majority of religious movements make little or no financial or material profit for their founders. More commonly, the religious innovators end up losing their fortune, family, and friends because of their claims. They end up losing social and financial capital that it took them years to earn. Their willingness to suffer abuse solely because of their views and their refusal to reconsider their claims when presented with overwhelming material incentives to do so suggests that they sincerely believe their claims.

Even though Ahmad's status as a feudal lord and the efficient British-Indian law and order machinery mostly protected him from physical abuse, he still suffered tremendously because of his claims. None of Ahmad's family members, who inhabited the same village as he, accepted his prophetic claims. They included his first wife and their children, his siblings, his cousins, and his uncles. His cousins were so strongly opposed to Ahmad that one of them called him "a cunning schemer" (Griswold, 1902: 25). Mirza Imad-ud-Din erected a wall to block Ahmad's access to their family mosque forcing Ahmad to go to court against Imad-ud-Din. Strong opposition by Ahmad's sons and cousins also played a critical role in preventing Ahmad from contracting his third marriage. Having claimed divine support for the marriage, Ahmad had to suffer significantly, when his own wife and son convinced parents of the young bride-to-be–Muhammadi Begum–to reject Ahmad's overtures. Anti-Ahmadi Muslims continue to use this episode in their polemic against the community to this day. One of Ahmad's friends and a prominent *Ahl-e-Hadith* leader, Muhammad Hussain Batalavi, who had earlier played a critical role in promoting Ahmad as a defender of Indian Islam, renounced his friendship once he heard of his prophetic claims. He swore to "take Ahmad down" (Dard, 1948). Batalavi traveled throughout India asking Muslim scholars to issue a fatwa of *kufr* (infidelity) against Ahmad. *Kafir* (i.e., an infidel) was not the worst name Muslims called Ahmad. Batalavi called his former friend by much worse names:

> raving drunkard, intriguer, swindler, accursed, the one-eyed Dajjal, slave of silver and gold, whose revelation is nothing but a seminal discharge, shameless, the ring-leader of sweepers and street vagabonds, dacoit, murderer, whose followers are scoundrels, villains, adulterers, and drunkards. (Dard, 1948: 575)

> That the Qadiani is a Dajjal of this time, a second Musaylimah, perfidious, deceiver, cheat, liar and impostor, and that he is the enemy of the faith of Islam and all other heavenly faiths. (Dard, 1948: 608)

According to Griswold, Batalavi was not alone in condemning Ahmad. Many North Indian Sunni Muslim leaders stood with Batalavi against Ahmad.

> In the numerous FATWAS, which Muhammadan Associations all over India have issued against the Mirza Sahib, the strongest words of denunciation are used. Thus he is called KAFIR 'unbeliever'; DAJJAL 'Anti-Christ', mulhid 'heretic', murtadd 'apostate', KAZZAB 'LIAR', Be-Iman 'Faithless', Daghabaz 'Deceitful' etc; etc; with such epithets as these is the 'certificate' filled, with which Muhammadan orthodoxy has dismissed the Mirza Sahib from its fellowship and service. (Griswold, 1902: 20)

In his 1898 book "Kitab-a-Bariya", it took Ahmad six pages to summarize the damning words used by his former friends and supporters among the Ahl-e-Hadith (Ahmad, 1898: 118-124). As a result of vitriolic anti-Ahmadiyya propaganda by Muslim leaders, on several occasions Ahmad was physically assaulted by mobs when he travelled outside his home village of Qadian. In 1905, while lecturing in Amritsar, he was pelted with stones (Ali, 1937: 53). When Griswold asked the aforementioned Indian Muslim about Ahmad in 1902, he was told that if Muslims such as "the Amir of Kabul were only in authority here" Ahmad would've lost "his head" (Griswold, 1902: 25). This was by no means the isolated opinion of a man on the street; Batalavi said the same thing.

> Had we been under Muslim rule, we would have given you (Ahmad) a proper reply. We would have at once cut off your head with a sword and made you a dead body. (Dard, 1948: 608)

Two of Ahmad's followers, who had the misfortune of living under the authority of the aforementioned Amir of Kabul, were jailed, tortured, and publicly stoned to death. Their families were exiled to Turkistan. At each successively brutal step in their torture, they were offered a chance to recant their faith in Ahmad but they refused choosing death over worldly gains.

If religious innovators are not mentally disturbed or charlatans, then why do healthy and seemingly rational human beings make the radical claims that they do? Stark and Bainbridge (1987) developed their *entrepreneurship model of religion* to answer such questions. They consider NRM founders to be entrepreneurs who produce, market, and sell *compensators* in exchange for other rewards (Stark & Bainbridge, 1987). A compensator is an unverifiable promise of a future reward that is in low supply or unavailable at present. According to the Stark-Bainbridge theory, in situations where some rewards are in low supply or not available at all, people are willing to accept compensators in lieu of the actual rewards. For instance, a religious founder may be able to sell the unverifiable promise of life after death to those agents that intensely value immortal life. Entrepreneurship theory's most useful contribution is highlighting the similarities between the role of an entrepreneur and an NRM founder who also has to fashion a new product, market it, and sell it. Elaborating these similarities has allowed the use of economic analysis tools for explaining the higher rates of religious participation in societies with more religious pluralism such as the United States as compared to religiously homogenous societies such as Sweden. However, the Stark-Bainbridge entrepreneurship model falls short of a complete theory as it does not explains as to why NRM founders invent and propagate new ideas that seem so radical to most of their fellow group members and why some people buy these ideas? In this book, I will reserve the term radical to refer to those ideas that are considered to be so outside the pale by the primary target audience members (invariably the group members whose interests the NRM founder is claiming to defend) that they consider the NRM founder to be a deviant and not a full-fledged member of their group.

The multidisciplinary approach I outlined in my 2005 article (Upal, 2005b) builds on Stark and Bainbridge's entrepreneurship theory as well as developments in the new field of *cognitive science of religion* (Barrett, 2000; Boyer, 1994; Lawson & McCauley, 1990; McCauley & Lawson, 2002), social identity theory (Tajfel & Turner, 1979; Tajfel & Turner, 1986; Turner, Hogg, Oakes, Reicher, & Wetherell, 1987), and leadership as social-identity-change-entrepreneurship (Haslam, Reicher, & Hopkins, 2005; Reicher, Haslam, & Platow, 2011; Upal, 2005b). Cognitive science of religion (CSR) assumes that ordinary cognitive processes result in creation and spread of religious ideas and that special mechanisms devoted to religious cognition are not needed to understand religion. Barrett calls it the naturalness-of-religion thesis.

> ...much of what is typically called 'religion' may be understood as the natural product of aggregated *ordinary* (emphasis in the original) cognitive processes. This perspective may be called the 'naturalness-of-religion thesis'. Much as language is naturally acquired as a result of cognitive preparedness plus exposure to a typical sociolinguistic environment, ordinary cognition plus exposure to an ordinary environment goes a long way towards explaining religion. (Barrett, 2000: 29)

The cognitive science of new religious movements which I proposed in 2005 subscribes to CSR's naturalness-of-religion thesis and focuses on connecting macro-level phenomenon of religion to micro-level cognitive processes. I assume that most NRM founders and believers are rational agents functioning in ordinary states of consciousness. The task for cognitive science of NRMs then is to identify the ordinary cognitive processes that cause NRM founders and believers to behave in ways that most observers find outlandish, and result in social ostracization of the NRM founders and their followers by the very community they claim to be saving. We will see that many of these processes are social psychological in nature. They are initiated and led by people who believe that their religious group's prosperity will be enhanced through a change in their shared beliefs and seek to sell that message of change to their fellow group members. Those community members who buy this idea become their followers and those who seek to preserve the old belief system become their primary opponents and chief persecutors.

As we will see in Chapter 4, I caution against the tendency among some cognitive scientists of religion who argue that universality of cognitive explanations blunts the need for understanding the sociocultural and historical context that leads to the creation and spread of radical religious ideas. Indeed, I strongly believe that an explanation of religion that focuses solely on universal cognitive processes that are common to all human beings is doomed to fail. Universal cognitive processes simply do not provide a complete explanation unless they are instantiated in a particular sociocultural context that is investigated by historians of religion. Thus a history of religion is (and always will be) a necessary complement for a cognitive science of religion. This also means that unlike most other cognitive science of religion works that you may have read, we will be doing a deep dive into the history of nineteenth century India to fully understand the context which led Mirza Ghulam Ahmad to develop his ideas that resulted in establishment of Ahmadiyya Muslim Jama'at and led people such as my maternal and paternal grandfathers to accept his innovative religious claims.

1.4 Sources for the Book

An interdisciplinary effort such as this must draw its sources from a variety of scholarly traditions. In the case of the cognitive science of NRMs theory and its application to understand the origin of the Ahmadiyya doctrine presented here, these include a history of South Asian Islam (particularly the 19th century), (Allen, 2006; Cox, 2002; B. Metcalfe, 1982; Powell, 1993), new religious movements research (Bainbridge & Stark, 1979; Stark & Bainbridge, 1987), cognitive science (Schank, 1999; Simon & Newell, 1972), social psychology (Hogg & Vaughan, 2002), and cognitive science of religion (Boyer, 1994).

For history and doctrine of Ahmadiyya Muslim Jama'at (AMJ or movement hereafter), I use non-Ahmadi (Western scholarship as well as Muslim) sources as well as Ahmadi sources. The early Westerners who wrote about AMJ were Christian missionaries working in India. These include Rev. Robert Clark (1825-1900), the first secretary of the Church Missionary Society who served in Punjab from 1851 till his death in 1900. Clark's adopted son Henry and Ahmad debated each other and following their debate Ahmad was charged by police for ordering one of his disciples to murder Clark (Ahmad was acquitted by the judge for lack of evidence). Missionary reports by Clark and others (such as Rev. Dr. Imad-ud-Din) published in the Church Missionary Intelligencer offer a rich source of not only of missionary perceptions of Ahmad and other Punjabi Muslims and their strategies for converting Muslims but also the broader socio-cultural environment in the nineteenth century Punjab. The earliest scholarly report on the Ahmadiyya movement was penned by the Presbyterian missionary Hervey De Witt Griswold (1860-1945), who moved to Jhansi, India in 1890. He seems to have met Ahmad around 1900. He wrote his report on AMJ in 1902 and presented it to the Victoria Institute in 1905 (three years before Ahmad's death). The American Methodist missionary, Rev. Howard Arnold Walter (1883-1918), published the first English-language book-length treatment of the movement in 1918. John Nicol Farquhar (1861-1929), a Scottish missionary to India, included a discussion of AMJ in his 1915 book, "Modern Religious Movements in India." The American Episcopalian missionary James Thayer Addison wrote an article on the movement for the Harvard Theological Review in 1929. More recent scholarly publications include the books by Spencer Lavan, Yohan Friedman, and Adil Hussain Khan and an article by Shazia Ahmad (S. Ahmad, 2010; Friedman, 1992; A. H. Khan, 2015; Lavan, 1974).

The primary Ahmadiyya source is the approximately fifty Urdu books written by Mirza Ghulam Ahmad as well as his letters ("Maktoobat") and pamphlets ("Ishtiharat"). These are available from the AMJ's official website (alislam.org) as well as the website run by the Ahmadiyya Anjuman Isha'at-e-Islam Lahore (aail.org). Official English translations of about forty of Ahmad's works are also available from alislam.org. Unless otherwise noted, citations from Urdu books are my own translation. The secondary sources include the five-volume "Malfoozat," a diary of his utterances noted by three of his disciples (Moulavi Abdul Karim, Mufti Muhammad Sadiq, and Yaqub Ali Irfani) and published in Ahmad's lifetime, and the eight of his *seerat* biographies also written by his disciples and published after Ahmad's death. The most extensive among these is the three-volume "Seerat-ul-Mahdi" written by his son Mirza Bashir Ahmad and published from 1927 to 1939. These are complemented by the twenty-one-volume history of the movement titled "Tarikh-e-Ahmadiyyat" written by the movement's official historian Dost Muhmmad Shahid and published by AMJ's official press (Shahid, 1958).

My final source of information about Ahmadiyya beliefs and practices is my own lifetime of "field observations" as a lay Ahmadi as well as part of the Ahmadiyya officialdom as I held various offices in the Jama'at organizations during the first twenty years of my life in Rabwah as well as immediately after arrival in Canada. By praying five times a day at the neighbourhood mosque and listening to the *daras* following the namaz, and by participating in Jalsa Salanas, Atfal and Khuddam Annual Ijtemas, Talim-ul-Quran classes, Ramadhan daras, and weekly mohalla meetings I learned Jama'at's version of its doctrine and history directly from Jama'at's best scholars. These include Sir Muhammad Zafrullah Khan, Dost Muhammad Shahid, Malak Saifur Rahman, Abdul Malik Khan, Abdus Sami Khan, and Hafiz Muzaffar Ahmad. I also had a chance to meet and listen to several aging companions of Mirza Ghulam Ahmad and Ahmad's grandchildren including the 3rd and 4th Ahmadi Khalifas, Mirza Nasir Ahmad and Mirza Tahir Ahmad, respectively.

1.5 Outline

The next three chapters lay out the multidisciplinary theoretical framework of the cognitive science of new religious movements. Chapters Five through Seven apply this model to Mirza Ghulam Ahmad's radical claims that were seen as outside the range of acceptable Muslim beliefs by many of Ahmad's Sunni Muslims peers whom Ahmad claimed to be defending against the vigorous Christian and Hindu missionary efforts.

Chapter 2 traces the origin of ideas of us-versus-them to our basic psychological need to feel good about ourselves. According to social psychologists, people need to have a positive self-esteem to function as healthy members of a society. The need to have a positive self-esteem drives various groups (including religious ones) to create myths that tell their members why they are better than members of their competitor groups. Because our cultural world is constantly changing, these myths need upkeep so that they stay relevant for each successive generation. This job is taken up by those high-status group members for whom group forms a large part of their identity. They have the most to lose if their group's status declines and most to gain if their group's status is elevated. Chapter 3 argues that NRM founders are such *social-identity-change entrepreneurs* who repair and refashion their group's myths ostensibly to ensure their group's future prosperity. It reviews the development of the ideas of social identity entrepreneurship from social psychology to politics to new religious movements.

Chapter 4 explains why people are naturally attracted to counterintuitive ideas. It starts with the traditional cognitive science of religion work on memory for counterintuitive religious concepts. It proceeds to present the context-based model of memory for counterintuitive ideas. By placing the cognitive science of religion work in the larger framework of psychological research on human memory, the context-based model explains that people are attracted to novel ideas because evolution favours eager learners who pay more attention to those aspects of their environment that their memory fails to predict. It also presents empirical evidence that I and my colleagues have collected through numerous lab studies of memory for various types of concepts. The evidence favours various predictions of the context-based model (Russell & Gobet, 2013). These include ratcheting up of counterintuitiveness that allows us to understand how layers of counterintuitiveness can build on top of one another and result in a complex tapestry of ideas that seems so bizarre to those who grew up outside of that tradition.

Unlike the traditional content-based model of memory for counterintuitive concepts, which asserts that there are differences between memory for *maturationally natural* ideas and *practiced natural* ideas such as the culturally-learned ideas, the context-based model asserts that there are no such differences. This allows us to generalize the traditional cognitive science of religion notion of individual counterintuiveness to *social counterintuitiveness* as violations of shared beliefs of a group of people. This key development, presented in Chapter 5, allows us to understand the spread of a much broader class of ideas including doctrinal innovations of new religious movement founders such as Mirza Ghulam Ahmad.

Chapter 6 focuses on the context of British Indian Islam in which Ahmad made his socially counterintuitive claims. An in-depth historical examination of sociocultural beliefs of 19th century North Indian Sunni Muslim elite allows us to uncover the shared beliefs of Ahmad's target audience and is critical to understanding the cognitive appeal of Ahmad's message for them. I review shared beliefs of North Indian Muslim elite about themselves, as well as about British government, Christian missionaries, and their eschatological beliefs about Jesus' second coming and about Mahdi. An understanding of a group's shared beliefs is crucial for figuring out their expectations, whether a concept is perceived as intuitive or counterintuitive by them, and whether or not they are able to make sense of the counterintuitiveness of a concept and hence perceive it as catchy.

Chapter 7 continues the historical study begun in Chapter 6 to examine the life of Mirza Ghulam Ahmad paying particular attention to religious influences in his life. It particularly explores the hitherto unexplored influence that nineteenth century reformist Ahl-e-Hadith leaders such as Syed Nazir Hussain and Muhammad Hussain Batalavi played in preparing the fertile ideological ground that gave birth to Ahmad and also led some among the Ahl-e-Hadith to make sense of his claims and become his early followers. I argue that historians of religion need to pay more attention to the Ahl-e-hadith context if they want to fully understand the origin of the doctrine of the Ahmadiyya Muslim Jama'at.

Chapter 8 traces the evolution of AMJ's doctrinal beliefs about Ahmad's prophetic claims in the period following Ahmad's death and leading to the split of the movement into two groups. Using the ratcheting-up-of-counterintuitiveness model, I carefully analyse the advantages and disadvantages of the competing strategies adopted by the two Ahmadi factions, namely, the Qadiani Ahmadi strategy of claiming a full prophetic status for Ahmad and the Lahori Ahmadi strategy of claiming a lower reformer status for Ahmad. I examine the attraction of each missionary strategy to their primary target audiences of fellow Ahmadis as well as non-Ahmadi Sunni Muslims whom both groups wished to convert to Ahmadiyyat. The chapter ends with an examination of the similarities between the branding strategies adopted by both factions in the West where an increasing number of Ahmadis have migrated to escape persecution in Pakistan and other Muslim-dominated countries, e.g. Bangladesh and Indonesia.

The final chapter (Chapter 9) concludes by summarizing the main axioms of the cognitive science of new religious movements and prospects for future work in this field.

2 Tribal Gods: My God Is Better than Yours

The mood among the courthouse clerical staff was jovial as they left work at the end of long day. A group of friends engaged in friendly banter. Somehow, the conversation turned to athletic prowess. One of the men, who happened to be a Sikh, boasted that no one could run faster than he. His 28 year old Muslim coworker took this to be a communal challenge to all Muslims and countered that he would be happy to race the Sikh man to the next bridge. A mutual friend of the two men volunteered to be the judge of the impromptu track competition. The judge ran ahead to the other side of the bridge so that he could determine the winner. Once there, he signaled for the competitors to start. The young Muslim challenger ran as fast as he could to save the honour of Islam. As he reached the end point ahead of the Sikh, the judge declared him to be the victor. Other Muslims gathered around him to congratulate him. They recalled that another famous 19th century martyr-saint, Syed Ismael,[14] who on hearing that a Sikh man is a great swimmer, took up swimming and after months of painstaking practice challenged the Sikh swimmer and beat him at his own game. That passion to save the honour of Islam led Syed Ismael to lead an army of Muslims against British and die fighting the colonial authorities and their allies (Allen, 2006).

The race took place outside the District Court House in Indian City of Sialkot in 1868. The young Muslim runner was Mirza Ghulam Ahmad, the future founder of Ahmadiyya Jama'at, who had left his home village just months ago to take up his first (and only) job in the large multicultural urban centre of Sialkot. Had he lived for another 79 years, Ahmad may have been happy to know that Sialkot (along with the rest of the Western Punjabi cities) was ethnically cleansed of all of its offending Hindu and Sikh inhabitants and had become part of the new Islamic country of Pakistan. One of Ahmad's ardent followers, Sir Muhammad Zafrullah Khan, became Pakistan's first foreign minister. Former Hindu and Sikh Sialkotis moved eastward to Indian Punjab, some of them settling in Ahmad's home village of Qadian which was left on the Indian side of the border. In one of the largest ethnic cleansings in the world history more than twelve million people were forced to abandon their houses, farms, friendships and in some cases blood relatives. What caused millions of people who had lived with their neighbours for generations to feel threatened and abandon everything they had owned and everyone they had known? Why is it so easy for "us" to become scared of "them"? Why did a young man take an innocuous little boast about athletic prowess to be a challenge to his religious faith? How are actions of partisan individuals such as Ahmad both defined by the intensely sectarian atmosphere in which they grow up and contribute to maintaining and intensifying it? How and why do defenders of faith such as Ahmad end up wanting to change the very religion they purport to defend? How do they end up being reviled by the very community they claim to be saving?

2.1 Wired for Tribal Thinking

According to social identity theory–one of the most well developed social science models of social behavior–having a positive self-esteem is one of the most deep-seated human desires (H. Tajfel & Turner, 1979; Henri Tajfel & Turner, 1986; Turner et al., 1987). Similar to the way, we need food and water to maintain our physical health; we need to feel that we are good people to maintain our mental health. One of the most amazing findings of the social psychological research is that our *identity* (i.e., our sense of who we are) is not defined just by our individual characteristics (such as the unique shape of our face) but also by the groups that we belong to. Not surprisingly then our *self-esteem* (i.e., what we think we are worth) is also based on our evaluation of the status of the groups we belong to. We feel happy when our group ranks well compared to other groups and sad when it does not seem to measure up. Just think of how you felt the last time your favourite sports team won a game. Perhaps, you felt the feeling of nirvana similar to the following fans of championship winning teams.

> I am so proud to call myself a Packer fan it is just not possible to put it into words. I thought the last Super Bowl win was special but it is not at all comparable to this! WAY TO GO GREEN BAY PACKERS!!!!! That includes players, coaches, management and especially fans! We are the best and I am very proud to be a part of it. (Comments on The Green Bay Packers forum)

As the following news report shows, American Football fans are not the only ones susceptible to such euphoric experiences following their team's victory.

> "Iniesta took us up into heaven" after an "agonising" game, said the sports daily Marca. "We suffered, but it was worth it."
>
> The country's prime minister, Mr. Zapatero said he was "happy and emotional".
>
> Fans were left pinching themselves after watching their team triumph.
>
> Said Adolfo, 25, said: "It's an extraordinary feeling of happiness and nerves."
>
> "We're going to celebrate like crazy, all night, all Monday until Tuesday morning," said 41-year-old Miguel Angel. (Winter, 2010)

Even if you are one of the unfortunate few who have not yet experienced the unmitigated joy and pleasure of basking in the glory of your team's championship win (sorry Winnipeg Jets, Texas Rangers, and Houston Astros fans), you probably know a die-hard sports fan who seems to be enjoying winning more than the players themselves. Sports psychologists have found that fans are more likely to wear their team's jersey after a win and claim credit for the win by using terms such as "we won."

Psychologist Edward Hurt of Indiana University found that ardent sports fans are more confident of winning dates with attractive members of the opposite sex following a win than after a loss. He showed pictures of attractive members of the opposite sex to sports fans whose teams had won and lost and found that fans of winning teams were significantly more confident of being able to obtain a date with persons shown in the pictures. Furthermore, he found that fans were also more confident of their ability to perform well in other unrelated activities such as darts and word games following a win by their team.

Bragging about the greatness of one's favourite team is a standard part of the sports-fan behavior repertoire. Other unseemly components of this repertoire include verbal arguments and physical violence between spectators at sporting events and these days also on social media. As the following excerpt from a blog report about a Philadelphia-New York Football game shows the quality of arguments between sports fans leaves a lot to be desired.

> With Sunday's 27-17 win over the Giants, the Eagles are easily and hands down the best team in the NFC and possibly all of football. Not the Jets, not the Patriots, not the Falcons, not the Steelers nor the Saints. The Philadelphia Eagles are the best team and the most dynamic as well. The game wasn't pretty by any accounts, but the Eagles have now shown that they can win by any fashion and how dangerous they are. (Nesgoda, 2010)

If you know any die-hard sports fans, you know that they are convinced that their team is the best, even if it is losing a game. New York Times reporter Warren St. John followed Alabama Crimson Tides fans as they followed their favourite college football team in their RVs from one tailgating party to another. However, instead of learning about the reasons for the blind devotion that fans have for their teams, he ends up confronting his own inner demons. Even though St. John was born and raised in Alabama, he had moved to New York when he was a teenager to study at Columbia. He is horrified to discover that he shares (with die hard Alabama fans) the belief that Alabama is better than other football teams. What's worse is that he cannot explain why he holds that belief.

> Try as I might, it's impossible for me to admit that I like Alabama "just because" and not due to an objective reality that establishes as more worthy of my affection than Tennessee or Auburn. (As I write this, a part of my psyche is battling for control of my finger tips and screaming out, 'Oh yes, we are!') But why? My rational mind asks. Because we cheat less? (Doubtful.) Because we have better colors? (Absolutely, but if tomorrow we changed to mauve I'd reluctantly go along.) Because we're better than they are? (Some years, yes; some years no - the historical tally is still being counted.) (St. John, 2005: 226)

I can relate to that. Whenever I have struggled with my faith in Ahmadiyya Islam, I have never considered any other Islamic traditions such as Sunni or Shia Islam, nor faiths of Zen Buddhism, or Jehovah's Witnesses (despite the fact that they insisted on dropping by our house in Saskatoon clutching a fresh issue of Watchtower magazine every month). Every time, I have asked myself 'Why not?', I feel a gut reaction because Ahmadiyya Islam is the better than any other kind of religion, and that somehow the Ahmadiyya belief system is more logical and more reasonable than others. I couldn't quite explain why but I just knew that Ahmadiyyat made more sense than all other religious belief systems. If I had to believe in an organized religion, it would have to be Ahmadiyyat. Off course, similar to St. John, I didn't have any rational argument to support my firm beliefs, yet, there it was, somehow lodged deeply inside the core of my being.

Social scientists have a name for such beliefs that are so deeply held that no evidence is considered necessary-*myths* (Lugli, 2013). Sports fans (both the closeted and the outted ones!) and religious zealots are not the only ones who believe in group superiority myths, many cultural groups hold similar beliefs. Anthropologists studying ethnic groups find that ethnocentric beliefs in "superiority of the ingroup's culture combined with condemnation of the outgroup as immoral and inferior" are "commonplace. 'Chosenness' is a particularly prominent expression of this belief" (LeVine & Cambell, 1972: 6). Political scientist, Stephen Van Evera argues that such chauvinist myths are "hallmark of nationalism, practiced by nearly all nationalists to some degree" (van Evera, 1994). He provides a number of illustrative examples including the Nazi myth of Aryan supremacy, British and American beliefs in rational and intellectual exceptionalism (Longley, 2003), and Russian belief in their extra-ordinary inventiveness (see Table 1 for these and a number of other group superiority myths from around the world). He could easily have added many more to his list such as the Pakistani belief that one Pakistani Muslim soldier can dominate 10 Indian Hindu soldiers, American Indian belief that they are more spiritual than the more material "white man", Israeli belief that they are more rational than crazy Arabs, Muslim belief that God chose to favour them as his final chosen people after Christians and Jews strayed from the prescribed path, and the Nation of Islam belief that an evil black scientist created the wicked white man. Group superiority myths are reflected in the literature and art of a group and feature prominently in its creation stories that form the master narrative of a group.

2.2 Muslim Hierarchy-Enhancing Myths

Although it's easy to see Christianity and Islam as vast and static forces, they are perpetually in flux. Over time each religion has shaped the other. (Griswold, 2010: 12)

Table 3: A sample of group superiority myths from around the world.

An Abkhaz Myth	When God divided up the Earth between nations, the Abkhaz were not present and were left without land. When the Abkhaz finally arrived, God asked them why they were late, and the Abkhaz replied: „Because we had guests and could not leave them to join you." God was touched by Abkhaz hospitality and decided to give them the land he had planned to keep for himself. (Bauldelaire & Lynch, 2009)
A Black Myth	Over eight thousand years ago, a brilliant but malevolent geneticist of the tribe of Shabazz named Yacub used his scientific knowledge to create the white man out of the original (Black) man by gradually mutating the latter's genes. (Marable & Agard-Jones, 2008:250)
A German Myth	We Germans are more humane than the other nations; we do have better blood and breeding, more soul, more heart, and more imagination. (Von Bulow, 1915: 46)
A British Myth	[British are] the greatest and most highly civilized people that ever the world saw" and are "the acknowledged leaders of the human race in the West (Van Evera, 1994: 48)
An American Myth	God... has made us the master organizers of the world... He has made us adept in government that we may administer government among save and senile peoples... He has marked the American as His chosen nation. (Beveridge, 1900)

Islam and Christianity are similar in many ways. Both claim over a billion adherents around the world. Both originated in the Middle East and both claim to be heirs to the Hebrew prophetic tradition starting with Abraham (Silverstein & Stroumsa, 2015). Both ascribe miracles to their founders: Jesus and Muhammad. Both Jesus and Muhammad were persecuted by those in authority. However, only Muhammad headed a state and led an army into battle. According to the Muslim tradition, the first major battle took place in 624 at Badr. Under Muhammad's leadership, just over three hundred ill-equipped ("naked" and "hungry"[15]) Muslims faced close to a thousand battle-ready Meccan *kuffar* (infidels) armed to the teeth with latest weapons and armor.

> The Holy Prophet (may peace be upon him) turned (his face) towards the Qibla Then he stretched his hands and began his supplication to his Lord:' Allah, bring about what Thou hast promised to me. O Allah, if this small band of Muslims is destroyed. Thou will not be worshipped on this earth.' He continued his supplication to his Lord, stretching his hands, facing the Qibla, until his mantle slipped down from his shoulders. So Abu Bakr came to him, picked up his mantle and put it on his shoulders. Then he embraced him from behind and said: Prophet of Allah, this prayer of yours to your Lord will suffice you, and He will fulfil for you what He has promised you. So Allah, the Glorious and Exalted, revealed (the Qur'anic verse): 'When ye appealed to your Lord for help, He responded to your call (saying): I will help you with one thousand angels coming in succession.' So Allah helped him with angels. (Sahih Muslim: Book 19, Number 4360)

The prophet grabbed some desert gravel and sand and threw it at the kuffar which caused a sandstorm blinding the kuffar. According to Muslims tradition, this is the incident referred to in Quran (8:15-18):

> O ye who believe! when ye meet the unbelievers in hostile array, never turn your backs to them.
>
> If any do turn his back to them on such a day - unless it be in a stratagem of war, or to retreat to a troop (of his own)- he draws on himself the wrath of Allah, and his abode is Hell,- an evil refuge (indeed)!
>
> It is not ye who slew them; it was Allah: when thou threwest (a handful of dust), it was not thy act, but Allah's: in order that He might test the Believers by a gracious trial from Himself: for Allah is He Who heareth and knoweth (all things).
>
> That, and also because Allah is He Who makes feeble the plans and stratagem of the Unbelievers. (Quran, 2000)

The infidels were routed. Seventy of them, including their leader, Abu Jihl, died on the battlefield while another seventy were taken hostage. The prophet, "looked at the people of the well (the well in which the bodies of the pagans killed in the Battle of Badr had been thrown) and said, "Have you found true what your Lord promised you?" Somebody said to him, "You are addressing dead people." He replied, "You do not hear better than they but they cannot reply." (Sahih Bukhari, Volume 2, Book 23, Number 452).

The dominant Muslim narrative of Badr is that with divine help Muslims were able to beat the odds and win over an enemy with vastly more material resources than them. According to Muslim tradition, this pattern repeated itself over and over again as Muslims quickly conquered Arabia and moved beyond its borders. Unlike the Meccan pagans, they were now up against Christians who also claimed to have God's final message giving them the divine right to rule. In what they saw a vindication of their claims, Muslims quickly rolled over their opposition taking one city after another in the Middle East and beyond: Bosra in 634, Damascus in 635, Jerusalem in 637, Khuzistan in 639, Alexandria in 641, Cairo in 642, Azerbaijan in 643, Fars, Kerman, Mekran, and Kharan in 644, Cyprus in 647, Kabul in 670, Samarkand in 677, Southern Spain in 711, and Sind in 712. The lightning fast advance firmly established the myth of Islamic exceptionalism. Most Muslims saw God's hand working behind the scenes to ensure their victories. The narrative that "if you follow Allah's path of Islam, you will always be victorious" became deeply embedded as a key component of Muslim identity (Upal 2015a; 2015b).

To vindicate his own claims to divine support for Christians, Pope Urban declared the first Crusade as a counter offensive against Muslims in 1095 (Philips, 2015). For the next three hundred years, the two civilizations militarily fought against each other in a broad front extending from Spain in the west to Syria in the east. Christians were able to expel Muslims from Spain, and Sicily while Muslim occupation of former

Christian lands of Middle East, North Africa, Turkey and the Balkans proved to be more durable. While Crusaders didn't succeed in their declared mission of retaking the Middle East, they did stop the lightning-fast expansion of the Muslim world at the expense of the mainly Christian West. The fatal blow against the Muslim expansion did not come from Christians. Instead, it came from the Mongols, led by Gengis Khan. The Mongol invasions of the thirteenth century were so ferocious that they broke through the Islamic narrative of invincibility. This was the first time that Muslims were forced to ask the question of what went wrong (Lewis, 2003). What reduced this cognitive dissonance for most Muslims was not a revision of the belief in God's promise of final victory of Islam over all other religions. Instead, as the 13th century cleric Ibn-e-Tamiyya explained, the reason Allah had taken victory away from Muslims was that they had stopped following Islam faithfully enough. The only way to restore the glory of the golden era of Islam was to go back to faithfully following Islamic tenets, in particular Jihad, argued Ibn-e-Tamiyya and others Muslim leaders.

While the Mongol rulers themselves eventually converted to Islam, and the Ottomans had some success in reuniting Muslims under one empire and making further inroads into Eastern Europe, Islamic countries never regained the degree of cultural and material dominance over Europe that they had enjoyed in the 11th and 12th centuries. The myriad reasons for this include the rise of science, reason, and free thinking in Europe and their relative decline in the Islamic countries.

3 Social Identity Change Entrepreneurs

> Late on the night of Friday June 1, 2001, Saeed Hotary left the Palestinian West Bank city of Kalkilya by car. Two colleagues, who also made the journey, dropped him off at the promenade in Tel Aviv. According to eyewitness reports, Hotary joined a long queue of people awaiting entry into the Dolphinarium nightclub. He mingled with some of the teenagers in the queue and flirted with one girl in particular. Without warning, Hotary detonated an explosive device strapped to his body, which held a large number of metal objects including ball bearings and screws. Within an instant, both Hotary and the girl evaporated. In total, twenty-one people died and one hundred and twenty were injured, the vast majority of whom were teenagers gathering on Tel Aviv's promenade to socialize at the weekend. (Gill, 2007: 143)

Social identity theory allows us to make sense of actions such as suicide bombings which are harder to understand using the traditional rational choice theory alone (although an integration of the two approaches is possible (Upal, 2014; Upal & Gibbon, 2014)). According to social identity theory, when people perceive their group's status to be lower than a comparison group's status, they are motivated to take restorative action to ameliorate the negative impact on their self-esteem, part of which is derived from their group membership. The social identity management strategies are typically classified into individual strategies and social strategies. Individual strategies include individuation and mobility. Individuation involves focusing on one's individual characteristics that distinguish one from other members of one's group. Depersonalization is the opposite of individuation where people come to see themselves as typical group members and indistinguishable from other group members. Since individual mobility strategy benefits the individual group members at the expense of their group, groups develop norms that view it negatively e.g., as selfish actions by greedy individuals. Depersonalization and social strategies on the other hand are viewed positively as heroic and unselfish. Group norms include mechanisms for rewarding those who lead social actions that favour the group.

Social strategies include collective violence as well as non-violent actions such as *social creativity*. Social creativity involves attempts to reshape shared social identity beliefs in favour of one's ingroup. These include choosing a different comparison group that has a lower status (so that the one's group looks better by comparison) and *dimension shifting* (elevating the importance of positive ingroup characteristics and downgrade those dimensions on which an outgroup looks better than the ingroup (Mummendey & Schreiber, 1984; van Knippenberg, 1978)) or outright attempts to denigrate the outgroup's status while boasting about one's ingroup. Psychologists argue that such ingroup favouritism and outgroup derogation naturally arise because of the group member's need to feel good about themselves. Since self-esteem of the individual group members depends on the perceived social value of their group, they are biased to preferentially process pieces of information such as rumors, jokes, and folktales that extol virtues of their group and denigrate other groups.

In a number of lab studies where subjects were arbitrarily assigned to groups (but told that they had some hidden characteristics in common with other group members), participants gave more rewards to members of their group than to members of other groups. In a series of experiments British psychologists, Henry Tajfel and his student John Turner, told 14-15 year old boys from a Bristol school that they had been placed in a group because of the way they estimated flashing dots or classified paintings. Then they were asked to assign points to other boys, some of whom had been placed in their group while others had been placed in a different group. They found that even arbitrary group labels such as "over-estimators" and "under-estimators" were sufficient to cause boys to give more points to boys in their group and fewer points to boys in the other group (Tajfel & Turner, 1979). If minimal groups set up in the lab where alleged group members may never get to see their fellow group members and which have no prior history, no cultural ties such as a common language, common dress, common food, common stories, or common poetry can control the behavior of its members, one can only imagine the power that real world cultural group must exert over their members.

Another dynamic that is hard to study in artificial groups is the degree of affiliation that members have for a group. In the real world groups, some members feel a stronger attachment to a group than do other members. These high identifiers are known to express more support for in-group favouritism beliefs than low identifiers. The sports fans who strongly identify with a team also report feeling happier than those who weakly identify with a winning team (Hirt & Clarkson, 2010). The more devoted fans are also more likely to report painful feelings such as despair, anger, and "being kicked in the gut" when their team loses, especially to an arch rival. Here's how St. John describes his feelings after his team's loss:

> when an undefeated Alabama team lost to Auburn my junior year, I anesthetized myself with a steady drip of keg Budweiser. The next morning I woke up on my dorm-room bed, fully clothed and in the fetal position. My roommates reported that I'd taken refuge there at some point in the fourth quarter and had wept myself to sleep...
>
> In the late spring of 1999, another team I'd foolishly adopted, the New York Knicks, lost in the play-offs. I should've been prepared—the Knicks always lost in the play-offs—but there I was again at the threshold of despair. I had a familiar internal dialogue: I blamed the players, then the coach, then the management, and of course the referees, and then I scolded myself for even bothering to care. (St. John, 2005: 13)

Members who identify strongly with a group are also more likely to search for causes of the events that led to a loss of their group's status. After convincing themselves of the reasons behind their group's subpar performance, they go about convincing others of these reasons, partly because they need other's support to "fix" the problems. Psychologist Dominic Packer found that Ohio State students who strongly identified with their school were more likely to challenge the perceived student binge-drinking norm than students who weakly identified with their school. The results were surprising because they question the traditional scholarly expectations that those who strongly identify with a group always want to maintain the status quo. The experimental results suggest that traditional wisdom is not true in this case. Once high-identifiers become convinced that the long term property of their group is in danger, they seek changes to the environments which may include changes to the group's cherished norms and shared beliefs.

The reason, why those who strongly identity with a group are more likely to challenge their group's shared beliefs seen to be threatening to the group, is that high identifiers have more to lose if their group loses its status and hence have more reason to be concerned about threats to its future status. High identifiers also have the social capital and the creds that are needed to withstand the blowback that always follows those who advocate changes to a group's norms.

Because social change, by definition, requires changes in beliefs and behaviors of a large number of people it is a painstakingly difficult process requiring years of efforts by highly motivated individuals. In order to successfully convince their fellow group members of the need for reforms, the reformers must have entrepreneurial qualities not too different from those required of used cars sales staff. They must work as *social-identity-change entrepreneurs* or SIEs for short. In order to be able to sustain their efforts and high motivation level over prolonged periods of time needed for social change, the SIEs come to see their purpose in life as spreading the message of change. They come to see themselves as social-identity-change entrepreneurs, as defenders of the group interests, and as reformers of ideologies be they religious, economic or political ideologies. This suggests that in order to affect changes to a group's social identity, SIEs must first affect a change to their own personal identity that allows them to start seeing themselves as SIEs.

3.1 Social-Identity-Change Entrepreneurs as Weavers and Sellers of Stories

> Evolution has "hardwired" the process and form of storytelling into human brains and mind. Fisher concludes that humans are really *homo narratus*, and that storytelling is an intrinsic human attribute. That is, story architecture is hard wired into the human mind. (Haven & Ducey, 2007: 11)

People around the world are avid storytellers. They love to tell and listen to stories. Compared to other activities such as reading and writing that require years of effortful training, listening to stories and understanding them seems to come naturally to people. A number of social scientists including the psychologist, Jerome Bruner, argued that the reason for that is that narrative processing is fundamental to our thinking. Bruner says that we have a "predisposition to organize experience into a narrative form, into plot structures and the rest" and that "we organize our experience and our memory of human happenings mainly in the form of narrative—stories, excuses, myths, reasons for doing and not doing, and so on" (Bruner, 1991: 4). We don't understand our lives as millions of unrelated random events but as logically connected sequence of events where everything happens for a reason. If someone asked you a question about a major event in your life, such as why did you do that, what would you say? Psychologists have found that most people tell stories in response. Here's the description of a social scientist who interviewed women in Kerala, South India.

> Asking a seemingly straightforward question (e.g. "what were the main causes of your separation"), I expected a list but, instead, got a "long story"... After coming to the end of a long and complex story of a marriage, a participant would sometimes say "uh, I'm afraid I got a little lost. What was the question you asked?"... Although my question ("were you ever pregnant?") could have been answered directly ("yes"), Gita chose instead to negotiate a space in the interview to develop a complex narrative. She describes terminated pregnancies, going to a political demonstration, coming home to her husband's anger, whereupon the scene shifts to the actions of in-laws' and her husband's refusal to be examined for infertility. (Reissman, 2001: 695)

Since our understanding of groups is an extension of our understanding of individuals, it is not surprising that we also think of groups as having lives with a sequence of events starting from how the group came to be and how it responded to various events in its environment including incidents involving other groups. Such narratives are shared among group members. In order to change their group's myths, social identity entrepreneurs not only have to understand the existing group narratives but similar to other marketers, they must also learn how to exploit them to create a narrative of change needed to sell their new social identity myths to their groups. This "selling with stories" technique is considered by marketers to be more effective than traditional techniques as marketing expert Scott McKee explains, "When I build a brand, I'm telling a story. When I have conversations, I'm telling and listening to stories. The narrative is the interesting part" (McKee, 2002).

A narrative of change must not only specify the problem with group's current thinking or behavior and the solution to that problem but it must also tell the group members why they should care about the problem and why they should adopt the solution being advocated. The narratives of change have to convince group members that current beliefs and behavior of the group members threatens the group's long term prosperity. They must also frame the offending beliefs and behavior of the group members as contrary to their identity as group members. They have to specify which of group's many beliefs and behavior is the true cause of the impending decline and they have to explain how the solution overcomes these shortcomings to ensure a glorious future for the group.

Social influence researchers have found that ardent group members, whose support is critical to any change effort, do not like messages that criticize the shared beliefs or behavior of their group. Similar to the way, St. John's psyche was struggling against his criticism of Crimson Tide fans, minds of high-identifiers counter-argue against any critique of their group. This makes them even more resistant to messages of change. One way to generate positive feelings in the minds of group members is by praising the group. Researchers have found, however, that simple addition of flattery to a message of change does not make such messages any more palatable (Packer 2008). Any group praise has to be seamlessly integrated with the message of change. A natural way to include praise of a group is to discuss the group's (real or imaginary) past glory and contrast it with the misery of the present state of decline (presumably due to the group member's offending beliefs and behaviors) and end the narrative with a promise of a bright future to usher in as a result of the adoption of the solution. Lehigh University social psychologists Dominic Packer, Gordon Moskowitz and Matthew Kugler conducted a series of studies to investigate the effectiveness of such social change messages (Upal, Packer, Moskowitz, & Kugler, 2011). They labeled social change messages that acknowledge that our group is currently not doing well, remind the group of its glorious past, and promise that making a change to group's shared beliefs will restore that glory in the future, as *arcing messages* (Figure 1). A series of studies found that such narrative structure was more persuasive than other structures especially on those group members that identify strongly with the group and are especially resistant to most social change messages (Upal, Packer, Moskowitz, & Kugler, 2011).

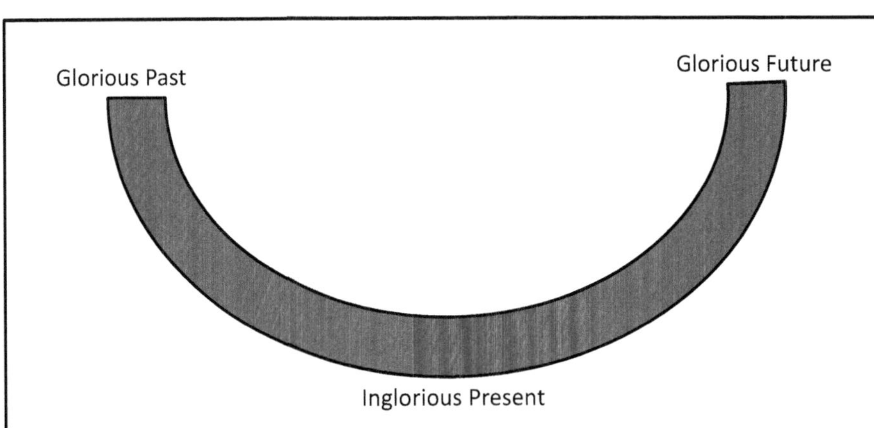

Fig. 1: Arcing pattern of glorious-past–inglorious-present–glorious future that successfully overcomes a group's natural resistance to messages of change.

Not surprisingly one can find instances of arcing messages in the speeches and writings of great leaders of all sorts of groups (such as professional, ethnic, and regional organizations as well civic, regional, and national governments). Because of their concern for the wellbeing and future prosperity of their group, such leaders are often able to spot issues that may impact their group's future prosperity and see opportunities that if pursued may lead to increased future prosperity of their group. Because of their strong attachment to their groups, they are willing to suffer the abuse of those who want to protect the status quo to seek desired changed in their group's shared beliefs. Such leaders perceive that their calling in life is to advocate the change in the group's shared beliefs. Because of their social identity change efforts such leaders can be called *social-identity-change entrepreneurs.*

While all US presidents worked for their nation's betterment (as they understood it), few modern presidents have been as eloquent advocates of their cause as President Regan. After getting elected, he wasted little time in selling his narrative of change. Here's part of President Regan's inaugural speech where he starts by identifying the problem:

> These United States are confronted with an economic affliction of great proportions. We suffer from the longest and one of the worst sustained inflations in our national history. It distorts our economic decisions, penalizes thrift, and crushes the struggling young and the fixed-income elderly alike. It threatens to shatter the lives of millions of our people. Idle industries have cast workers into unemployment, human misery, and personal indignity. Those who do work are denied a fair return for their labor by a tax system which penalizes successful achievement and keeps us from maintaining full productivity. (Committee-on-Inaugural-Ceremonies, 1989 :332)

Next he presents his diagnosis:

> For decades we have piled deficit upon deficit, mortgaging our future and our children's future for the temporary convenience of the present. To continue this long trend is to guarantee tremendous social, cultural, political, and economic upheavals. You and I, as individuals, can, by borrowing, live beyond our means, but for only a limited period of time. Why, then, should we think that collectively, as a nation, we're not bound by that same limitation? (Committee-on-Inaugural-Ceremonies, 1989:332)

The solution by now is obvious but President spells its out anyways.

> It is time to check and reverse the growth of government, which shows signs of having grown beyond the consent of the governed. It is my intention to curb the size and influence of the Federal establishment and to demand recognition of the distinction between the powers granted to the Federal Government and those reserved to the States or to the people. All of us need to be reminded that the Federal Government did not create the States; the States created the Federal Government. Now, so there will be no misunderstanding, it's not my intention to do away with government. It is rather to make it work -- work with us, not over us; to stand by our side, not ride on our back. Government can and must provide opportunity, not smother it; foster productivity, not stifle it. (Committee-on-Inaugural-Ceremonies, 1989 :333)

Selling the solution involves more than just presenting it. One of the problems that reformers face is that by presenting the dangers to the community, they end up alienating the strongly identifying group members who are naturally inclined to question every negative characterization of their group. In order to overcome group member's resistance, social identity entrepreneurs must often balance negative characterization of the group's present with overly positive characterization of its past and future and emphasize the strength of their affiliation for the group. They often argue that they criticize the group because they love it. In a survey of anti-Iraq war blogs, Packer (2008) found that a vast majority couched their opposition to the war in terms of their love for their country. Here's how President Reagan tries to soothe the feelings of his audience for his criticism of US government in his inaugural speech.

> If we look to the answer as to why for so many years we achieved so much, prospered as no other people on Earth, it was because here in this land we unleashed the energy and individual genius of man to a greater extent than has ever been done before... Directly in front of me, the monument to a monumental man, George Washington, father of our country. A man of humility who came to greatness reluctantly. He led America out of revolutionary victory into infant nationhood. Off to one side, the stately memorial to Thomas Jefferson. The Declaration of Independence flames with his eloquence. And then, beyond the Reflecting Pool, the dignified columns of the Lincoln Memorial. Whoever would understand in his heart the meaning of America will find it in the life of Abraham Lincoln. (Committee-on-Inaugural-Ceremonies, 1989 :333)

He was even more articulate about how adopting his solution of reducing the size of government would lead to a bright future in his 1984 State of the Union speech.

> America's best days and democracy's best days lie ahead. We're a powerful force for good. With faith and courage, we can perform great deeds and take freedom's next step. And we will. We will carry on the tradition of a good and worthy people who have brought light where there was darkness, warmth where there was cold, medicine where there was disease, food where there was hunger, and peace where there was only bloodshed. (Committee-on-Inaugural-Ceremonies, 1989 :333)

The esteem in which President Reagan is held by Americans, especially by Republicans, shows us the effectiveness of the arcing message of social change. New York's mayor Rudy Guiliani follows the same template when he identifies the 'liberal cynicism' as the root cause of New York City's problems. In his inaugural speech in 1994, he said:

> The era of fear has had a long enough reign. The period of doubt has run its course. As of this moment, the expressions of cynicism – New York is not governable, New York is not manageable, New York is not worth it – all of these I declare politically incorrect. (Guiliani, 1994)

He knows that he needs to sell his diagnosis of the problem to New Yorkers. He says:

> it's time for us to convince the cynics that the fear, the doubts and the cynicisms are over with... Don't let those who are so fearful of transformation stop the process before it begins. Killing ideas by fear. We don't need to be fearful. (Guiliani, 1994)

The solution to this problem, argues Guiliani is to overcome cynicism and doubt and renew and expand New York's traditional strengths. This he promised would lead to a glorious future for New York.

> Look anew at Broadway, the opera, the ballet, the museums and the fashion industry as powerful magnets drawing people and commerce to our city. We must expand them all. We must build our future on institutions like this. And as they grow, more New Yorkers will go back to work and have jobs... New York City has the very best health-care institutions in America. But we're not thought of and promoted as America's health-care capital. It should be so and it will be so. We're the home of over 100 institutions of higher learning in every field. We have some of the greatest schools in the world here in New York City. People come here from all over the world to be educated in our schools. Yet we don't think of New York as a college and university town. It should be so and it will be so... Albany, the capital of New York State. Washington, D.C., the capital of the nation. And New York City will again be the capital of the world. (Guiliani, 1994)

Mayor Giuliani's social-identity transformation of New Yorkers is widely seen as successful. During the first two years in office, New York's serious crime rate had dropped by half while the number of murders was reduced by a third. By the end of his two terms, the city's crime rate had fallen by 57%, so much lower as compared to other US cities, that New York was named as America's safest large city by the FBI.

Despite calling themselves conservatives, those on the right are constantly redefining their social-identity, from Joe Six Pack to Joe the Plumber and from Soccer Moms to Hockey Moms or rather a constant parade of social-identity entrepreneurs such as Reagan, Giuliani, and Donald Trump are at work to sell new ways of being authentic to a seemingly endlessly hungry public that keeps buying them. A February 2010 Rasmussen poll shows that 73% Americans believe that Washington is broken. This is remarkable for a number of reasons. The top reason being that this was a little over a year after Americans had just elected yet another "outsider" Barack Obama as President to fix Washington! Indeed, it seems that for decades at election time, almost every candidate seems to run "to fix Washington" yet it remains broken! Why is that? Why are social groups constantly in need of reformation? The issue is not specific to the American conservative movement or even to Western societies or even modern societies.

No human culture, past or present, seems to be immune to change and evolution. Shared beliefs systems from every domain whether it be politics, religion, art, or science appear to be extremely dynamic. Part of this is because of the role of social identity change entrepreneurs. Physical, social, and cultural environments inhabited by groups are constantly changing. Since reformed cultural groups are better groups (in the sense of being better suited to their environment) and therefore, groups have mechanisms that reward successful reformers. The SIEs get privileged access to resources such as money and mates and are therefore driven to seek reforms. However, this on its own is not sufficient to explain the level and pattern of cultural dynamism. Cultural trends are simply too numerous and too frequent. They also follow a peculiar pattern. The new cultural trend has a paradoxical relationship with the previous trend. It is defined with respect to it and also defined by an explicit opposition to it. The next chapter will look at the attraction that groups have for new trends and new ideas that I argue drives some of the cultural dynamism.

4 Attraction of the New

> While I was attending college in 1991, a friend of mine named Ghulam Hussain came to me and suggested I meet an Ahmadi Muslim friend of his. Ghulam felt Ahmadiyya views were very odd and thought I would get a good laugh out of hearing what they believed. I was actually excited because I had never spoken to an Ahmadi Muslim and had only heard the usual anti-Ahmadiyya propaganda that one hears in Pakistan.
>
> I went to the house of a man named Rashid Rabbani in Lahore. Rashid had a pleasing disposition and had his own business in Cant, providing surgical instruments to hospitals. I chatted with Rashid for about an hour and asked him a lot of basic questions: Why do Muslims consider Ahmadi Muslims to be "non-Muslim"; what is the Ahmadiyya view of the advent of the Imam Mahdi (guided leader) and Promised Messiah; and what is the Ahmadiyya view of the 'Seal of Prophethood'. The explanations Rashid gave me were intriguing, so before leaving I asked for some references. He gave me some pamphlets providing Qur'anic verses which I took and went home to review. A week went later, I went back to Rashid's house and told him that the references he had given me were indeed correct and I was interested in learning more about Ahmadiyya teachings... (Hussain, 2009: 122-123)

Syed Shahahzad Hussain's conversion story is not unique. I heard similar stories dozens of times from various new Ahmadis. Since Ahmadis were a small minority in almost every village, town, and city of Pakistan (except my hometown Rabwah of course), the Ahmadi message was perceived as unorthodox everywhere in Pakistan. The idea that Jesus had lived to an old age, died a natural death (as opposed to having been raised to heaven) and was buried in Kashmir and that Mirza Ghulam Ahmad was a prophet ran counter to the orthodox Sunni Muslim doctrine. Not surprisingly thus, Ahmadis are considered to be non-Muslim by many non-Ahmadi Pakistanis. Ahmad, the founder the Ahmadiyya Movement, was accused of being a false prophet and *dajjal*–a colorful anti-Christ-like supernatural villain who is supposed to ride a fire-driven donkey that can jump thousands of miles in one leap. It was widely rumored that Ahmadis stealthily say Ahmad under their breath instead of saying Muhammad when they recite the Muslim Shahada (creed) "there is no God but Allah and Muhammad is his messenger." Furthermore, Ahmadis were said to pray facing towards Ahmad's hometown of Qadian in India instead of praying towards Mecca-Muhammad's hometown. Whenever I traveled outside Rabwah, non-Ahmadis would ask me what the man-made Ahmadi heaven of *Bahishti Maqbara* (we'll discuss this special cemetery at length in Chapter 5) looked like? How many virgin *houris* (virgin girls that Muslims believe will be provided to male inhabitants of paradise) does it have?

Ahmadis are not the only social group that espouses beliefs that appear to be at odds with those from outside the bubble. As psychologist Jonathan Baron observes, for better or worse, such beliefs are widespread among a variety of social groups around the globe.

> ...it is amazing that various organized groups adhere to such bizarre versions of the truth as the following:
>
> Memories of "satanic ritual abuse" are frequently repressed and recovered in psychotherapy.
>
> Aliens from other planets come to earth and rape people.
>
> The Holocaust was a hoax.
>
> The U.S. government is planning to bomb its own citizens in order to establish a New World Order of international government.
>
> Blacks are not fully human.
>
> H.I.V., the AIDS virus, was produced by scientists and disseminated through Black neighborhoods for the purpose of genocide.
>
> The end of civilization is imminent. (Baron, 1998: 886)

The radical messages of religious innovators such as Ahmad offend the orthodox and lead to all kinds of punishment for those who adopt the unorthodox beliefs. The unorthodox community has to suffer violent financial, political, and social consequences for years. Hussain's conversion story suggests that there may also be some advantages to having an unorthodox set of beliefs. In this chapter we examine this idea in more detail. Could it be that the novel ideas of religious innovators that so offend the orthodox, also end up attracting new converts to their message? We explore this interesting possibility in depth in this chapter.

4.1 Is Religion Counterintuitive?

In 2001, the anthropologist Pascal Boyer published the book, "Religion Explained," which propounded a novel hypothesis. He wrote:

> ...religious representations are particular combinations of mental representations that satisfy two conditions. First, the religious concepts violate certain expectations from ontological categories. Second, they preserve other expectations. (Boyer, 2001: 62)

Boyer argued that most religious concepts are "minimally counterintuitive" (or MCI for short) because they violate a small number of category expectations while preserving most expectations. Minimally counterintuitive concepts such as *a talking tree* can be contrasted with intuitive concepts, such as *a green tree*, that do not violate any expectations and *maximally counterintuitive concepts* (or MXCI for short) that violate a large number of category expectations (such as *a talking tree that floats in the air and disappears at noon*). Boyer said that the reason minimally counterintuitive ideas come to be embedded in the minds of a population is that they enjoy some transmission advantages over other types of concepts. He further hypothesized that minimally counterintuitive ideas are better remembered than intuitive and maximally counterintuitive ideas. Boyer (Boyer and Ramble 2001) and other cognitive scientists of religion (Barrett and Nyhoff 2001) designed experiments to test this hypothesis. They gave subjects a number of short stories containing various types of concepts and asked them to carefully read them so that they could "answer a few questions about them." For instance, the following story was used by Barrett and Nyhoff (2001).

The Journey Home

A girl and a boy, sister and brother, were walking home from school on an ordinary day in an ordinary town. As they were walking towards home, they came upon a dog belonging to one of their friends. The cat crouched on the front lawn as it composed a symphony. Since the cat was completely absorbed in its work, the two siblings continued on their way, chatting about what they had learned in school that day, until a beautiful rose jumped right in front of their path.

The children knew that they must be getting home, or else their mother would begin to worry, so they slipped away from the rose. They had hardly made any progress in their journey, when they noticed an earthworm crossing the sidewalk. The girl picked it up and felt its slimy texture and the squirm of its movement. She placed it on the grass on the other side of the pavement and they continued on their way.

As they were walking, the boy's shoes sprouted roots which broke up the pavement below and impaired his movement. He had had this trouble with these shoes several times before and always carried a pocket knife in order to cut the roots. (Barrett & Nyhoff, 2001: 97).

The above story contains intuitive concepts such as "a slimy squirming earthworm" and minimally counterintuitive concepts such as "a jumping rose," "a cat composing a symphony," and a "root sprouting shoe." When people were asked to recall the stories, Boyer and Ramble (2001) as well as Barrett and Nyhoff (2001) found that people did indeed remember minimally counterintuitive ideas significantly better than intuitive ideas. They did not study recall for maximally counterintuitive concepts.

In 2002, while causally browsing through a "new arrivals" book shelf at a Barnes and Nobles, I picked up Boyer's book, "Religion Explained" (because of its surprising title, I have to admit!) and couldn't put it down until I had finished it. I was in the dog-house for several days for ignoring my wife and my family's weekend plans. What fascinated me about Boyer's approach was that unlike traditional sociological approaches to the study of religion which explain one social factor, namely formation of a religious movement, in terms of other social factors such as socio-political and economic factors, it purported to explain a social phenomenon in terms of individual cognitive factors. Now if cognitive scientists succeed in explaining how cognitive structures and processes could be implemented using brain's neural network hardware, and if neuroscientists could explain how chemical interactions between various brain chemicals can account for the working of the neural networks then we would have a truly natural science of religion!

My initial excitement about Boyer's theory soon gave way to some doubts about the power of the minimal counterintuitiveness (MC) hypothesis to explain the spread of religious concepts in the real world. In particular, I wasn't convinced that some of the most widespread religious concepts, namely, the Jewish-Christian-Islamic concepts of God and ghosts were minimally counterintuitive. To me, they seemed to be maximally counterintuitive, yet they were some of the most successful religious concepts. How did such religious concepts out-compete minimally counterintuitive concepts to become lodged in the minds of millions of human beings for centuries? Some cognitive scientists of religion have responded to this objection by suggesting that this is because believers hold two different ("theologically correct" and "intuitive") conceptualizations of God and that only the intuitive conceptualizations enjoy the transmission advantages because they are the only ones that are minimally counterintuitive. Cognitive Scientist of religion, Justin L. Barrett, says:

> God, and perhaps other religious objects and entities, are conceptualized on at least two different levels: the basic, everyday concept used in real-time processing of information, and the Theologically Correct (TC) level used in theological discussion of God's properties or activities outside of a real-time context. As was shown in above, these two levels of conceptualization may represent God in substantially different ways. (Barrett, 1997: 59)

Fellow cognitive scientist of religion, Jason Slone, who refers to intuitive cognition as "online" and TC cognition as "offline," explains the difference between the two ways of thinking as follows.

> Try this. Stand up and look down at your feet...
>
> What caused you to do this? Did God? Or did you do this by your own free will? Most likely, your instinctual answer is that you did it on your own. One reflection, you might postulate that it was all part of 'God's plan." If so, your cognitive efforts would be in line with how many religious people think. Your instinctive, or "online," answer is that you did it. Yet, if you believe that God is in control, then you might, on "offline" reflection, change your mind: God made me do it.
>
> Online thinking involves rapid judgments about things without much reflective thought. Offline thinking is more slowed-down and reflective. As such, offline thinking allows individuals to draw on learned schema...
>
> ... online thinking involves the employment of non-cultural, probably realistic schemas. No one had to teach you that if you want to stand up you have to make a choice to do it then act on that choice. You are hardwired to know that from birth. In this sense, we have a strong sense of what we might call "self-agency." (Slone, 2002: 115-116)

Thus, argue these cognitive scientists of religion, that the MC-hypothesis "does not apply" to the doctrinal conceptualizations of God or to any other cultural concepts that do not involve violating expectations of intuitive reflective thinking. This includes ideas that have been learned through explicit training such as the socio-cultural and religious schemas, scripts, and scientific concepts that people acquire after having been immersed in a socio-cultural or religious community or through exposure to long periods of training and practice such as the knowledge acquired by chess experts or scientist i.e., the so called "maturationally natural concepts". Maturationally natural concepts according to Barrett are "a natural product of human maturation in ordinary human environments... acquiring these competencies occurs without special artefacts or explicit tuition." Barrett's reason for focusing on these concepts is that they are "not the product of particular cultural conditions, but may be regarded as pan-human. Hence, it plays a role in all thought and communication regardless of cultural participants" (Barrett 2008: 311). Culturally- learned concepts, Barrett et al. argue, "will not provide an explanation for cross-culturally prevalent classes of concepts" (Barrett & Nyhoff 2001: 71). This is as opposed to the "practiced natural" knowledge which is acquired:

> with right sort of training or practice... consider the skills of a chess master or the performance of a concert violinist... this sort of naturalness requires special artefacts (e.g., a chess set or violin) typically involves explicit instruction, and varies considerably within groups (few become chess masters) and across groups (some cultures have more chess masters than others). Cultural schemata and scripts (e.g., how one orders food in a restaurant) may acquire this practiced naturalness in people. (Barrett, 2008: 310-311)

He argues that it is only the maturationally natural knowledge and not the practiced natural knowledge that provides the intuitive expectations whose violation leads to preferential processing for the counterintuitive concepts.

Another hurdle in the applicability of the MC-hypothesis to the spread of cultural beliefs in contemporary social groups is the implicit assumption by some that the MC-hypothesis is only applicable post-hoc to explaining how religious ideas arose in the evolutionary past but has little to tell us about how religious ideas are likely to evolve now or in the future. Dennett (2006), for instance, restricts his discussion of the MC hypothesis to the section on "evolution of religion" and does not mention it in the section on "religion today" in his book *Breaking the spell: Religion as a natural phenomenon.*

4.2 Just for Laughs

Canadian Broadcasting Corporation's "Just for Laugh Gags" is one of my favourite candid-camera TV shows (Just-for-Laughs, 2015a). I love how they set up elaborate hoaxes on the streets of Montreal to trap unsuspecting citizens going about their daily business. For instance, on one of the shows, unsuspecting park visitors look visibly puzzled when the dog starts going “baa baa baa…” like a sheep (Just-for-Laughs, 2015b). As the startled prank victim looks for the source of the sheep sound, the dog starts neighing like a horse or mooing like a cow. In another gag from the same episode, people look equally if not even more puzzled when a food store cashier proceeds to eat customer’s ice-cream and drink their coke instead of ringing their purchases into the cash machine (Just-for-Laughs, 2015c). Since the first gag involves a violation of the maturational naturalness (namely, a dog making horse or sheep sounds) and hence would be labeled counterintuitive by traditional cognitive scientists of religion but the second gag would not be considered counterintuitive because it only involves a violation of practiced naturalness. However, having watched enough episodes of Just for Laugh Gags, I did not have any reason to believe that the two concepts were treated differently by people. It didn’t make sense to me to restrict the label counterintuitive to violations of maturational naturalness. I thought that cognitive scientists of religion were being overly restrictive by limiting the minimal counterintuitiveness hypothesis to the spread of maturationally natural concepts in the evolutionary past. Given my graduate training and research on concept learning and memory, I was naturally curious about the learning and memory processes that led to differences in the way “maturationally natural” and “practiced natural” ideas were learned and represented as hypothesized by cognitive scientists of religion. Since the cognitive science of religion literature didn’t specify what these processes were and since I didn’t recall reading about such distinctions in the cognitive psychology and artificial intelligence literature during my graduate training, I decided to have another look at recent work in cognitive science of concept learning and broaden my search to the larger psychological literature on learning and memory. In the rest of this chapter, I’ll provide you a summary of what I found and the conclusions that I reached, some of which question the key assumptions made by some of my fellow cognitive scientists of religion.

4.3 Semantic Memory

Try this exercise, read the following number: 18649234

Next, read the following story:

Hockey Battle of the Sexes

In 1865, a Mormon man traded his nine cows for two wives. He has four sons and three daughters. He was very happy because he had exactly the right number to have hockey’s battle of the sexes. His four neighbours act as linesmen and referees.

Now turn the page.

Do you remember the number you saw on the last page?

Here's a hint, it had eight digits in it.

Still, no dice?

Don't feel bad. One of the most robust findings in psychology is that most people find it difficult to remember more than 7 unrelated pieces of information without constantly rehearsing them. This is known as the capacity of the *short term memory*. Short term memory being the transitional part of memory where information is kept before it can be integrated into the much larger long-term memory.

How about the story you read on the last page? If you are like most people you remember most of it (probably not word for word but the gist of it). Why is that? Why do we have such a hard time remembering an 8-character-number but can easily remember a much longer story? (over 200 characters in this instance) This is opposite of how a traditional computer-memory works. Clearly, human memories are organized very differently from a computer's memory. While computer memories are stored and retrieved by the address of the memory cell where they are stored, human memories organize information by their semantic relationships. Thus walking through the woods and hearing an owl hoot rhythmically can remind me of hearing the sound during my childhood during my visits to my father's home village in Sialkot, Pakistan, and of my grandmother, and the potato-filled naan-bread she used to make for us. In order to maintain a semantic organization, considerable cognitive effort has to be expended to find possible relationships between a new piece of information arriving from the senses and the semantic knowledge one has so that the new piece of information is stored with the information that it is most relevant and/or similar to. This *categorization* process allows us to assign meanings to new information by relating it with what we already know. Thus, when you walk into an office and see an object with four legs, a seat, and a back, your brain relates it to hundreds, perhaps even thousands, of such objects that you may have observed in the past and categorizes it as a '*chair*.' In addition to allowing us to maintain a semantically organized memory, categorization also allows us to remember important features of a new object and forget its unimportant features (such as scratches on its arms and the color of the tacks under its feet).

Since you did so well in the last experiment, it's time to conduct another one. Please read the following statements and indicate which ones are true and which ones are false.

A canary can sing.

A canary has blood.

A sparrow is a bird.

An ostrich is a bird.

Cognitive psychologists have found that people answer the first question much faster than the second question (Anderson, 2005). In general, scientists have found that people are quicker at determining the features that are specific to a subcategory such as canary than features that members of the subcategory share with members of the larger category such as animals. This led scientists to hypothesize that semantic memory is organized in a hierarchical fashion with general concepts represented higher and more specific concepts represented lower in the hierarchy as shown in Figure 2.

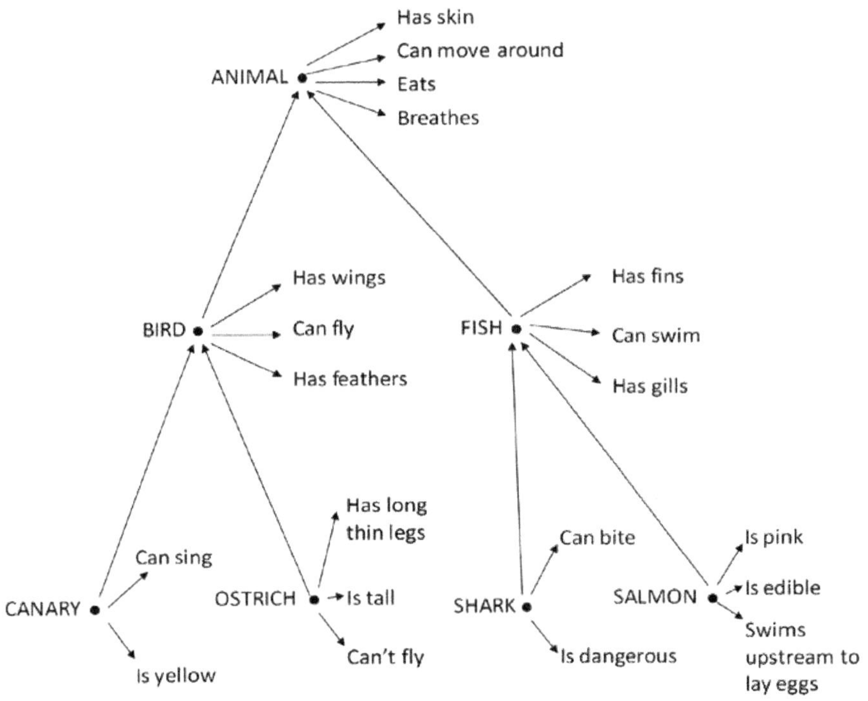

Fig. 2: A hierarchical concept view of some animal concepts.

The hierarchical view explains that when a person comes across a bird that it classifies as a canary, it is easily able to access features of the category canary. In order to answer questions about properties of the super-categories such as bird and animal, the memory system must search up the semantic memory tree, which takes longer. The problem with the hierarchical view is that it has difficulty accounting for the finding that even at the same hierarchical level, some information (e.g., sparrow has feathers) is confirmed faster than some other information (e.g., canary has feathers).

Ok, time to try another experiment. This time your task is to specify whether a given sequence of letters forms a valid English word or not.

Ardvark
bread
butter

Psychologists (Myer & Schvaneveldt, 1971) have found that people are faster at confirming that butter is a word when it's preceded by the word "bread" rather than being preceded by the word "nurse." Notice that butter and bread do not have a feature-category or super ordinate-subordinate category relationship. The only connection between the two words is that we often see the two objects together in the kitchen and we often hear the two words together. This suggests a non-hierarchical memory organization where concepts are connected to each other strongly if they occur frequently together and connected together weakly if they only occur together rarely. They do not have a connection if they are never, or almost never, seen or heard together. Such a memory organization is known as a spreading activation model of semantic memory.

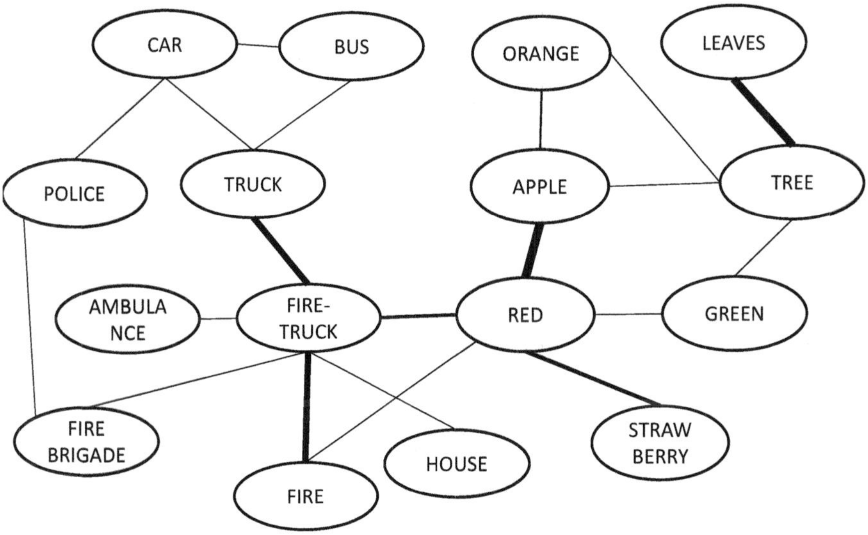

Fig. 3: An example of a spreading activation semantic memory model. The graphical representation shows concepts as nodes and connections beteeen nodes as edges. The size of an edge represents the strength of the connection between two concepts.

In a spreading activation model, concepts are connected to each other with varying degrees of strength as shown in Figure 3. Thus the concepts of TRUCK[16] and FIRE are more strongly related to the concept of FIREETRUCK than concepts of AMBULANCE and HOUSE. The spreading activation model suggests that when a concept is processed and is activated, the activation spreads out to its immediate neighbours as a function of the strengths of their mutual connections. For example, activation of the concept FIRE-TRUCK will strongly activate the FIRE concept while activation of the FIRE node will only weakly activate the concept RED. Spreading activation has been employed to explain the results of priming experiments that asked people to judge whether a sequence of letters formed a word. People were faster at this verification if they were asked to verify two related word.

The spreading activation model can explain the memory priming effects discussed earlier by hypothesizing a strong link between a stimulus and a prime. When the stimulus (e.g., "bread") is activated, it passes some of its activation to the prime word (e.g., "butter") which induces the expectation that the prime will be observed next. If the following word was expected then it is easier to confirm than if the following word was unexpected.

4.3.1 Learning a Spreading Activation Memory

A number of spreading-activation models (Anderson, 1983; Myers, O'Brien, Balota, & Toyofuku, 1984) couple a spreading-activation memory with a learning theory. While exact details of learning algorithms employed by different spreading activation models differ, general idea is that as objects or events are perceived to co-occur in an environment, and hence their corresponding nodes activated, the strength of their interconnections increases. If an object or event is seen in isolation, its links with its unseen neighbouring concepts get weaker. Thus overtime, an agent's memory comes to reflect the perceived frequency of co-occurrence of different objects and events not just in the "real world" comprised of objects and events but in the larger world of information which also includes cultural folktales, literature, news stories, movies, and social media. Thus if an agent lives in a world where donkeys are the primary mode of transportation, the donkey concept would become strongly connected to the ride concept. There appears to be considerable empirical evidence suggesting that the frequencies with which concepts have occurred together in the past are strongly predictive of their occurring together in the future (Schooler, 1993; Schooler & Anderson, 1997), and people are sensitive to co-occurrence information (Ellis, 2006). Thus Ellis (2006) concludes that:

> The words that we are likely to hear next, their most likely senses, the linguistic constructions we are most likely to utter next, the syllables we are likely to hear next, the graphemes we are likely to read next, and the rest of what is coming next across all levels of language representation, are made readily available to us by our language processing systems. Not only do we know the constructions that are most likely to be of overall relevance. . . but we also predict the ones that are going to pertain in a particular context. (Ellis, 2006: 7)

Since a concept, such as a bird, and its features, such as feathers, co-occur frequently, a concept will have strong connections with its features in an agent's mental models. Thus when a concept such as a bird is activated, the concept of feathers also becomes activated in such an agent's mind. However, if the agent learns that the bird in question does not have any feathers then its expectation is violated and a learning opportunity presents itself. In order to benefit from this opportunity, however, the agent must be able to justify that the lack of feathers does not preclude the animal from being a bird. If this justification process (which I have also called the *postdiction process*[17] to emphasize the fact that it is triggered after a concept and its attributes have been seen results in a coherent concept then that concept is remembered well (Upal, 2005a). The concepts that simply cannot be made sense of are not remembered as well. This is a rational thing to do in a world where an agent's sensors are not perfect and where sometimes things do happen more or less randomly.

The postdiction process is a crucial component of the context-based model. It can employ an agent's prior world knowledge as well as the knowledge provided to it in the context in which the concepts are presented. For instance, the concept of non-flying bird may be part of a story in which the bird lost its wings in an unfortunate hunting accident. If the justification process allows the agent to establish a coherent concept of a bird-without-wings then that concept will be added to the agent's long term memory and remembered well. In my 2005 paper (Upal, 2005a), I argued that evolution should favour the development of cognitive architectures that devote memory resources to those concepts that promise the most information gain to a learning agent. I further hypothesized about two types of fast and frugal heuristics (Gigerenzer, Todd, & Group, 1999) that people may be using to estimate information gain. I suggested that given a concept, people use their estimate of:
- **predictability:** the degree to which people expect to see that object or event in the near future and
- **postdictability:** the degree to which the expectation violation can be made coherent within the context in which it occurs.

Predictability of a concept *C* can be measured by stopping readers just before they read the concept in question, and asking them "how likely is it that the next concept in this story is *C*". This is similar to the process used by educational psychologists to measure cloze probabilities for various words (Bloom & Fischler, 1980). These probabilities are used to measure reading difficulty of a piece of text. Postdictability of a concept can be measured by allowing people to read the full piece of text all the way to the end and then asking them about the amount of effort it took to make sense of that concept. For my 2005 study, I used two questions to measure postdictability of various types of concepts. The first question asked people who had read the journey home story, "what would 'a jumping flower' mean." The follow up question asked them to rate the difficulty of answering this question on a 1-10 scale where 10 was labeled "impossible" and 1 "very easy." The answer to the follow-up question was used to measure postdictability of various concepts.

The context-based model suggests that people use the difference between postdictability and predictability (postdictability – predictability) to estimate the likely information gain. Thus concepts that are surprising (thus have a low predictability) should be better remembered as long as they can be made sense of (i.e., they have a high postdictability). Thus intuitive concepts are not remembered well because they have high predictability and high postdictability and therefore they offer little information gain to a learner. The maximally counterintuitive concepts are also not recalled well but for the opposite reasons, namely, because they have low predictability as well as low postdictability. Their low predictability indicates to the reader that there is a gap in the reader's world model, however, the reader is ill equipped to take advantage of this learning opportunity because she does not possess enough knowledge needed to make sense of it. Thus MXCI concepts are too difficult for their audience (it's a bit like trying to teach quantum mechanics to kindergartners). As Figure 4 shows, minimally counterintuitive concepts fall in the sweet spot of the concepts that have low predictability but high postdictability. Thus they are harder to predict and indicate a gap in the reader's knowledge and the reader possesses just the knowledge needed to make sense of them. Thus like MXCI concepts, MCI concepts are surprising but unlike MXCI concepts (and similar to INT concepts), they are learnable.

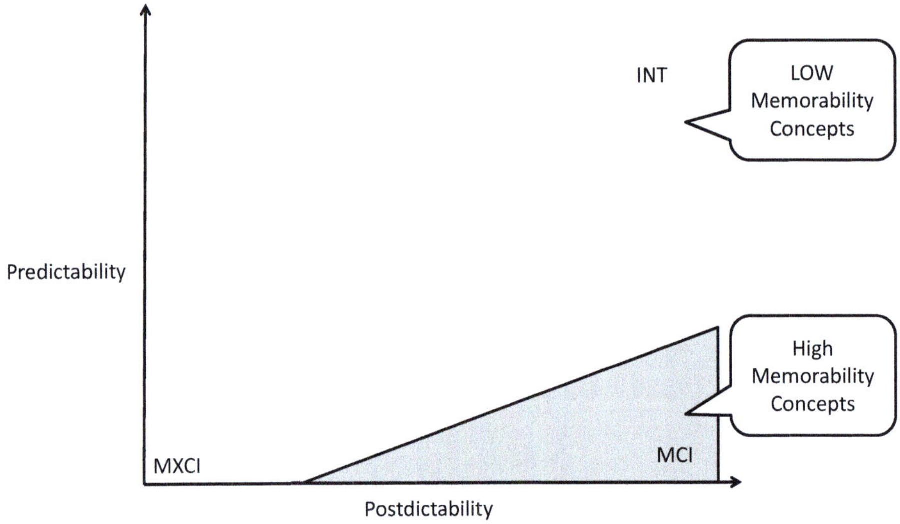

Fig. 4: The context-based model explain variation between predictability, postdictability, and memorability of intuitive (INT), minimally counterintuitive (MCI) and maximally counterintuitive (MXCI) concepts as the difference between their predictability and postdictability values.

To understand the context-based model better consider the following story (a version of which was used in experiments reported in (Upal, 2009)).

Obscurity Brings Safety

Once, a man, who was invisible, ran into a woman who could see invisible objects. The all-seeing woman said what is a beautiful man like you doing being invisible. Were you visible, no maiden could refuse you. You are missing out on all the fun. On hearing this, the invisible man decided to have his body painted with skin color so that people could see him. On his way home from the paint shop he was mugged and wished that he had remained invisible as obscurity brings safety.

When readers read the concept of a man, it activates their mental concept of MAN which activates all the concepts that are strongly connected to it including the concept of having a physical body which can be seen. However, upon finding out that the man is invisible, their expectations are violated and they engage in a justification process to explain reasons for this expectation violation. The readers may reason that this story belongs to the genre of moral fables[18] and use their world knowledge about fables to infer that fables often employ supernatural characters to illustrate a useful truth[19]. Furthermore, the readers can justify the man's invisibility as needed to support the story's plot which supports the moral lesson (and story's title) "obscurity brings safety". Since, this successful justification process accesses the conceptual structures in the agent's long term memory and thereby activates them, the story's title may become strongly connected to the story's plot and through it to the INVISIBLE-MAN node. This means that subsequently when readers are presented the title of the story, some of the activation will be passed to the INVISIBLE-MAN concept making it likely that it will be recalled by the readers.

Contrast this with the following story of similar length, with the same title, and containing the same number of MCI concepts:

Obscurity Brings Safety

Once a man who had feet instead of hands ran into a woman who was made of iron. The iron-woman said what is a beautiful man like you doing being difficult? Were you not difficult, no maiden could refuse you. You are missing out on all the fun. On hearing this the man with four feet decided to have his body painted with skin color to become more attractive. On his way home from the paint shop he was mugged and wished that he had not done that as obscurity brings safety.

In this story, although the reader's expectation about a person having only two feet is violated (indeed, this is also an MCI concept of the same variety as the concept employed in the first version of the story seen earlier), a reader may not be able to quickly construct a justification for this violation even in the context of a fable since the expectation-violation is not helpful for illustrating the story's moral lesson. This means that the concept of MAN-WITH-FOUR-FEET will not be recalled as well as the concept of INVISIBLE-MAN. This is what we found in experiments we carried out to measure differences in recall between counterintuitive concepts that were easier or harder to justify in the context of the story. Participants in our study recalled those concepts whose inclusion could be easily justified more frequently than those concepts which were harder to postdict and therefore harder to make coherent.

The idea of conceptual coherence has been increasingly emphasized by concept acquisition researchers in cognitive (Murphy & Medin, 1985) and developmental psychology (Gopnik & Meltzoff, 1997) as critical to understanding how people acquire concepts in the real world. Unlike the traditional feature-similarity based accounts of concept acquisition, which categorize an object or event based on how similar its features are to those of existing category members, the newer "theory-based" models of concept acquisition focus on the role played by the concept learner's background knowledge in grouping seemingly dis-similar items into a coherent conceptual category. As Murphy and Medin point out, this formulation of concept acquisition differs from that of naturalness (Keil, 1981) on which Boyer and Barrett base their traditional content-based account of the minimal counterintuitiveness effect.

> It is important to distinguish this notion of coherence from the related one of *naturalness* as used by Keil (1981) and others. Natural concepts are said to be formed out of basic ontological categories, such as living thing or intelligent being. For example, a category that included only thoughts and fish would cross ontological boundaries improperly and would therefore form an unnatural concept. However, as we later show, a concept that is unnatural (according to this definition) may be coherent because people have some theory that it plays a part in. (Murphy & Medin, 1985: 291)

Murphy and Medin go on to explain how unnatural concepts such as the Biblical concepts of *clean animals* and *unclean animals* and the category *prime-numbers-or-apples* can be made coherent and thereby remembered and recalled well by a concept learner.

> Suppose that one of our colleagues in the math department, Wilma, has only two interests: prime numbers and apple farming. We might, then, form the concept *prime-numbers-or-apples*... By adding more explanatory links, one could make the concept more coherent. For example, one could try to explain why Wilma has only those two interests. Through reference to naive personality theory and by exploring properties of apples and prime numbers, one could elaborate a theory about why a person would have just these interests. If this theory were consistent with one's other world knowledge, then it would also supply external structure to the concept. (Murphy & Medin, 1985: 298)

The context-based model posits that counterintuitiveness is a property of the context in which a concept appears[20] as much as it is a property of the concept itself. This can be contrasted with Barrett's content-based view which downplays the role played by context and defines counterintuitiveness as a property of the concepts alone. Barrett & Nyhoff (2001) claim that this is necessary in order to explain cross-cultural success of religious concepts.

> Although research on schemas and scripts suggests the possibility that incongruent concepts may be better remembered thus contributing to their transmission, these conceptual structures are culturally variable to a large extent and will not provide an explanation for cross-culturally prevalent classes of concepts. (Barrett & Nyhoff, 2001:71)

Thus according to Barrett, the goal for cognitive science of religion is to fully explain religion through universal cognitive processes that are common to all human beings in all times thereby avoiding the messy process of trying to understand historical and sociocultural contexts in which religious ideas arose. The problem is that this overly limits the potential application of cognitive science of religion models leaving it unable to explain much of what makes religion interesting. For instance, if we ignore the role that context plays in making concepts counterintuitive and memorable, we cannot understand how maximally counterintuitive concepts central to many world religions, such as Judo-Christian and Islamic concept of God, came to be widespread. It also leaves us unable to understand how radical claims by new religious movement founders such as Ahmad attract converts. The context-based model, on the other hand, sees historical and sociocultural contexts as crucial to understanding how concepts were seen as counterintuitive by people who were attracted to them. Thus cognitive science of religion I envision sees historical and sociocultural approaches to the study of religion as necessary complements to a cognitive account of religion.

The context-based model that I developed in 2005 (Upal, 2005a) was also helpful in understanding sometimes seemingly contradictory results of the existing studies of memory for intuitive and counterintuitive concepts (Atran, 2004; Barrett & Nyhof, 2001; Boyer & Ramble, 2001). Thus I argued that Atran's (2004) participants, who were presented lists of concepts, may have been able to recall intuitive concepts better than MCI concepts because they were not provided the context of a story (thus giving them lower postdictability). However, a scientific theory should not only be able to explain past observations but also make predictions which can then be tested through experiments. I carefully developed various consequences of the context-based model and teamed up with my colleagues with expertise in cognitive psychology (Professor Ryan Tweney of Bowling Green State University and Professor Mary Harmon-Vukic of Providence College) and religious studies (Professor Jason Slone of Georgia Southern University) to empirically test these predictions.

4.3.2 Teasing Apart Consequences of the Content-Based and Context-Based Model

Both content and context-based models lie within the general epidemiology of beliefs framework developed by cognitive anthropologist Dan Sperber (Sperber, 1996). Sperber's model focuses on exploring competition between ideas to understand why certain representations become widespread and cultural while others do not. Both approaches also assume that better recall for minimally counterintuitive concepts is a result of evolutionary processes that result in people having a memory architecture that causes minimally counterintuitive concepts to be better remembered than other types of concepts. However, the two approaches differ in a number of different ways and have different consequences. We illustrate some of these differences next.

4.3.2.1 Memory for a Concept Is Affected by the Context in which It Is Placed

Do you remember the story of a boy and girl walking home from school we saw at the beginning of this chapter entitled "The Journey Home"? It had some counterintuitive concepts such as "a cat composing a symphony", "a jumping rose", and "shoes sprouting roots." What would happen if we change the context in which these concepts are embedded to read as follows?

> **My Dream**
>
> I have always been fascinated by dreams. I have always wondered as to why some of our dreams are so different from our everyday experience; why are the laws of nature are violated so often in our dreams even though that never happens when we are awake? I remember that in some of my childhood dreams, our puppy Jack would to talk to me. In others, I would fly through the air. However, this particular dream must have been more bizarre than others because I wrote it down in my diary when I woke up.
>
> July 7, 1985. Last night, I dreamed that my sister and I were walking home from school. We saw a cat belonging to one of our friends. The cat crouched on the front lawn as she composed a symphony.
>
> Since the cat was completely absorbed in her work, we continued on our way, chatting about what we had learned in school that day, until a beautiful rose jumped right in front of us. We knew that we must get home soon before our mother starts worrying so we slipped away from the rose.
>
> We had hardly moved when we noticed an earthworm crossing the sidewalk. I picked it up and felt its slimy texture and the squirm of its movement. I put it on the grass on the other side of the pavement and we continued on our way. As we were walking, my sister's shoes sprouted roots which broke up the pavement below and impaired her movement. I remembered that she had had this trouble with these shoes before and carried a pocket knife in order to cut the roots.

Would the three counterintuitive concepts be better remembered in the new context of "my dream" or the old context of "the journey home"? While the content-based model does not provide any guidance on how to answer such questions, the context-based model specifies that we can compare concept memory in various contexts by comparing the concept's predictability and postdictability in each context. For instance, most participants in our 2007 study (Upal, Gonce, Tweney, & Slone, 2007) found the counterintuitive concepts to be significantly more predictable in the "my dream" version of the story i.e., people expected a child's dreams to have more counterintuitive concepts in it than the story describing an actual event in an ordinary town on an ordinary day.

Assuming that there's no significant difference in postdictabity of the counterintuitive concepts in the two stories (i.e., the counterintuitive concepts can be made sense of with equal ease in both stories) we can predict that the counterintuitive concepts should be recalled better in "the journey home" story context than in the "my dream" story. In 2005, we gave some undergraduate students at a Northwest Ohio university "my dream" version of the story while others only saw the "journey home" version. When both groups of students were asked to recall the stories they had read, a significantly larger number recalled the counterintuitive concepts embedded in "the journey home" version just as predicted by the context-based model. This shows that by changing the context, we were able to change memorability for counterintuitive concepts.

The context is not limited to the sequence of words before and after a concept, but includes a broad set of contextual conditions including the background knowledge that the readers possesses prior to learning the new information (Pazzani, 1991), the reader's motivation (Guthrie, Wigfield, Metsala, & Cox, 1999), and the resources (such as time) available (Heit, 1998) to comprehend the information being presented to them. Changing, any or all of these contextual factors can affect a concept's memorability and different concepts may be more or less memorable for different people in different situations.

4.3.2.2 All MCI Concepts Are Not Created Equal

Do you find the following two concepts equally plausible or do you find one of them to be more plausible than the other one?

- There is a someone who can hear prayers from billions of people from around the world at the same time, and
- There is someone who has a human body and a jackal head.

Cognitive scientists of religion label both of them as "minimally counterintuitive" (MCI) because each concept only violates one of our categorical expectations about people. However, for someone who has grown up in a Judeo-Christian or Islamic culture, the first concept may not be as surprising as the second concept. On the other hand, for ancient Egyptian believers in the cult of Anubis, the second concept may be less surprising. Since not all concepts assigned by cognitive scientists of religion to a category (INT, MCI, or MXCI) have the same predictability and postdictability, and since memory for a concepts is a function of its predictability and postdictability, not all INT, MCI, and MXCI concepts should have the same memorability. Some INT concepts may be better remembered than other INT concepts. Similarly, some MCI concepts may be better recalled than other MCI concepts and some MXCI concepts may be better recalled than other MXCI concepts. Furthermore, some INT concepts may even be better remembered than some MCI (or MXCI) concepts. Some MXCI concepts may be better remembered than some INT or even some MCI concepts. This means that there is no hard boundary between INT, MCI, and MXCI concepts in terms of memorability. Thus we can find an MXCI concept, an INT concept, and an MCI concept such that the MCI concept is less memorable than the INT and the MXCI concept, even though on on average we know that MCI concepts have better memory than INT and MXCI cocepts.

4.3.2.3 Intuitive Expectation-Sets

Responding to a growing series of studies criticizing the traditional content-based model (Gonce, Upal, Slone, & Tweney, 2006; Tweney, Upal, Gonce, Slone, & Edwards, 2006; Upal, 2005a, 2007a, 2007b; Upal et al., 2007), Barrett presented a revised version of the model in 2008 (Barrett 2008). The revised content-based model divides people's intuitive ontological knowledge into five categories of "Spatial Entities, Solid Objects, Living Things that do not appear to be self-propelled, Animates, and Persons" (Barrett, 2008: 317). Barrett labeled the expectations that people form upon hearing of a member of a category as an *intuitive expectation set* for that category. He provided a table (reproduced as Table 4) describing the five intuitive expectation sets for the above categories.

Table 4: Barrett's Intuitive Expectation Sets.

Category	Properties
Spatiality	Specifiable location in space and time
Physicality	Cohesion (move as connected whole) Contact (physical contact required for launching or changing direction of movement) Continuity (movement is continuous in space) Solidity (cannot pass through or be passed through by other solid objects) Tangibility Visibility
Biology	Growth & development Like begets like Natural composition Nourishment needs and processes to satisfy those needs) Parts serve the whole to sustain life Vulnerability to injury & death (if animate, seeks to avoid injury & death) Kind-specific essence
Animacy	Goals "Self-propelled" (including moving in space, changing appearance, emitting sounds, etc.)
Mentality	Reflective & representational mental states (e.g., beliefs, desires) and standard relationships among them and limitations of them (e.g., limited perceptual access) Self-awareness (including emotions and epistemic states) Understand language & communication

The context-based model (Gonce et al., 2006; Tweney et al., 2006; Upal, 2005a, 2007a, 2007b; Upal et al., 2007) views human semantic memory as a network of concepts with their connection strengths encoding their semantic overlap. According to this view, a category is strongly connected to its features because they frequently co-occur. Thus when a category label is activated (e.g., when an agent observes an object), its associated features will also be activated leading the agent to expect the object to have those features. A "domain" in the context-based view thus is a set of propositions that are strongly connected in the agent's semantic memory and may or may not perfectly correspond to Table 4.

People's expectations can be elicited by asking them appropriate questions. A number of feature-norming researchers (Ashcraft, 1978; McRae, Cree, Seidenberg, & McNorgan, 2005; Vinson & Vigliocco, 2002) have formulated such questions. McRae et al. (2005) conducted the largest feature norming study to date, to understand what features people expect when they encounter one of the 541 living and non-living things such as dog and chair. The participants were provided 10 blank lines to list features for each concept. They were asked to list physical (perceptual) properties, functional properties, and "other facts about it, such as the category it belongs in or other encyclopedic facts (such as where it is from)" taking as much time as they needed (McRae, Cree, Seidenberg, & McNorgan, 2005). Unfortunately, none of the previous feature-norming studies included the very high level concepts of solid objects, and mental beings that cognitive scientists of religion believe are crucial to understanding the spread of religious ideas. In a study reported in Upal (2014c). I created four high level concepts of solid objects, living things, animals, and mental beings to correspond to Barrett's domains of physicality, biology, animacy, and mentality. Replicating McRae et al. (2005)'s methodology, I asked 250 English speaking participants recruited through Amazon's Mechanical Turk service what features they expected of each of these four categories.

The features our participants most commonly listed for the concept "a solid object" are graphically shown in Figure 5. The percentage of participants who listed a feature is indicated as the strength of the connection between that feature and the concept in question. Only half of the features from Barrett's list—namely *tangibility* ("can be touched," mentioned by 14% of the participants), *visibility* ("is visible," mentioned by 9%), and *solidity* ("is solid," mentioned by 7%)—made the list of features mentioned by our participants; *cohesion*, *contact*, and *continuity* were not mentioned at all. Furthermore, the three features most commonly mentioned by our participants were not included by Barrett in his expectation set. These are *hardness*, *heaviness*, and *having a mass*.

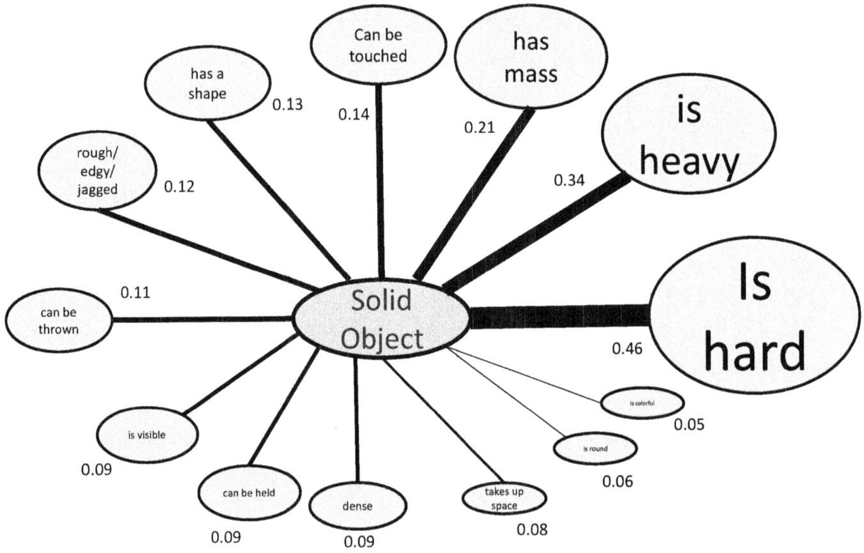

Fig. 5: Most commonly listed features of the category "a solid object" mentioned by by participants in Upal (2014c). The number beside each feature node represents the proportion of participants who mentioned the given feature.

The features our participants most commonly listed for the concept "a living thing" are graphically shown in Figure 6. The results show that our participants endorsed four of the features mentioned by Barrett (2008), namely, *nourishment* ("can eat," mentioned by 25% of participants), *reproduction* ("can reproduce," mentioned by 22%), *growth* ("can grow," mentioned by 14%), and *vulnerability* ("can die," mentioned by 11%). None of the participants mentioned *natural composition*, *kind-specific essence*, or *parts-serving-the-whole* features. In addition, the most commonly mentioned feature of *breathing* was not included in Barrett (2008). This may be because not all living things actually breathe, but that is beside the point. Work on naïve/folk psychology has shown that people's intuitive expectations do not always correspond to scientifically defendable notions. Other less frequently mentioned features, also missing from Barrett's table, include *moves/walks/runs* (19% participants) and *is alive* (16% participants).

Figure 7 shows the features most commonly listed by our participants for the concept "an animal." The results show that while a significant number of participants (30%) mentioned some type of movement (*walking*, *running*, or *moving*), only one participant explicitly listed *self-propelled* ("moves on its own") as a key feature of being an animal. The second feature listed by Barrett (2008)—namely, *goals*—was not mentioned by any of our participants (nor were any of its synonyms: *purpose*, *objective*, or *aim* listed). On the other hand, the features most commonly mentioned by our participants—namely, "eats food" (63%), "is alive" (47%), and "breathes" (32%)—are not included by Barrett in Table 1 as key features of *animacy* (although *nourishment* is listed as a key feature of *biology*, a superordinate category of *animacy*).

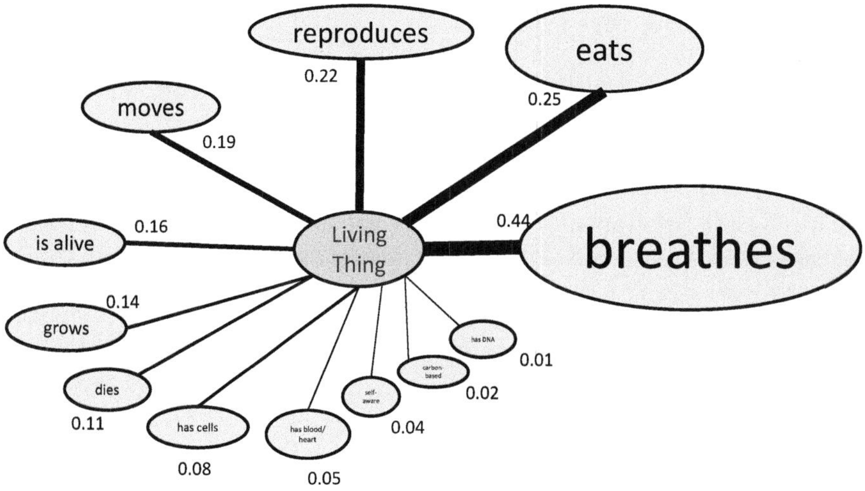

Fig. 6: Most commonly mentioned features of the category "a living thing" by participants in Upal (2014c).

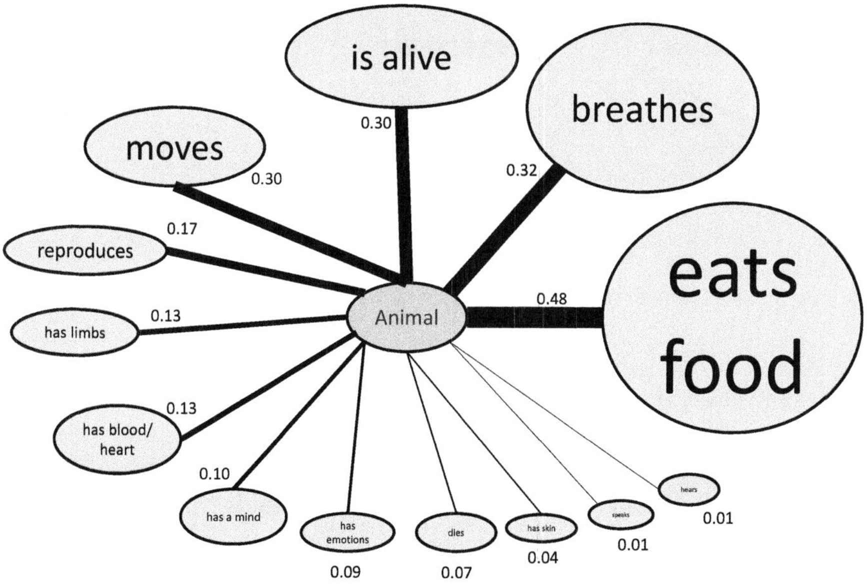

Fig. 7: Features most commonly listed for the category "an animal" by participants in Upal (2014c).

The features mentioned most commonly by our participants for the concept “mental being” are shown in Figure 8. It shows that only a small number of the participants listed the three features mentioned by Barrett:

> *self-awareness* (“is self aware/conscious”), mentioned by 8% of participants;
>
> *understand language and communication* (“can perceive” and “can talk,” mentioned by 9% and 8% of participants, respectively); and
>
> *reflective and representational mental states* (“has beliefs/desires”), mentioned by 6% of participants.

Furthermore, none of the top features mentioned by our participants—“thinks” (72%), “is human” (23%), “is an animal” (19%), “has emotions” (19%), “is smart” (15%), “is alive” (13%), and “has a mind” (12%)—is included by Barrett (2008).

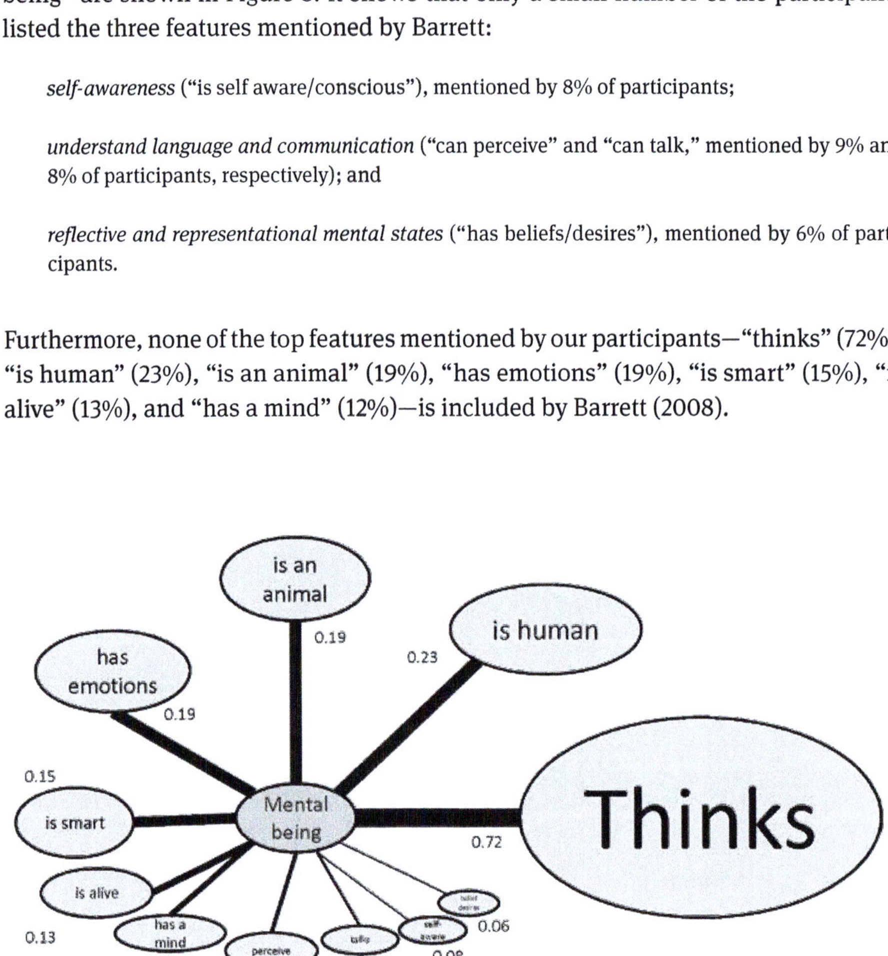

Fig. 8: most prevalent features of the category “a mental being” by participants in Upal (2014c).

While a majority of participants agreed that mental beings think, even the most prevalent features for the categories “a solid object” and “living thing” and “an animal” were mentioned by only a minority of participants. These findings can be used to revise Barrett’s intuitive expectation set shown in Table 5.

Table 5: Revised intuitive expectation sets based on studies reported in Upal (2014c).

Concept	Properties
Solid objects	Are hard, rigid, and firm.
	Are heavy (i.e., they have a weight).
	Have a mass.
	Are tangible (can be touched, have a shape, can be thrown).
	Are visible (they can be seen).
Living things	Breathe.
	Eat food/nourish themselves.
	Reproduce (like begets like).
	Are able to move.
	Grow and develop.
	Are vulnerable to injury and death.
Animals	Have limbs (hands and legs).
	Have blood and a heart.
	Have a mind.
	Have emotions.
Mental beings	Think (reason/reflect/have thoughts).
	Are human.
	Are animals.
	Are smart/intelligent/rational.
	Can perceive the world.
	Are self-aware/conscious.
	Talk to other mental beings.
	Understand language and communication.

4.3.2.4 Some Intuitive Concepts Are More Suprising Than Some Countrintuitive Concepts

The revised content-based model implicitly acknowledges the role of context when it suggests that multiple transfers from the same intuitive expectation set do not increase the counterintuitiveness score of a concept, a true appreciation of the role of context would go further and admit the possibility that some multiple transfers may actually make a concept *more coherent* and therefore more memorable than a single violation of the same concept. For instance, the concept of a "rock that listens to people, empathizes with them and helps them" may be more coherent when embedded in a story where the rock lies on a leaky roof of a poor amputee living in isolation than a rock that simply listens. This is because in a spreading activation semantic memory the concepts activated by an intuitive expectation set reinforce each other resulting in a more coherent and therefore more memorable set of ideas.

To test these predictions of the context-based view we created intuitive and counterintuitive concepts using Barrett's categories and then paired them with one and two properties from a domain to create four types of statements (Upal 2015e):

1. CE: Category label + one intuitive property.
2. CEE: Category label + two intuitive properties
3. CC: Category label + one counterintuitive property.
4. CCC: Category label + two counterintuitive properties.

We then asked people to indicate their degree of agreement or disagreement with each of the resulting statements on a 7-point scale. We found that as predicted by the context-based view statements with two intuitive properties were considered by people to be the less surprising than statements with one intuitive property which were considered less surprising than counterintuitive statements. Contradictory to the content-based view, people ranked the statement with two counterintuitive properties to be less surprising than statements with one counterintuitive property. According to context-based view, this happens because mental activation of one counterintuitive property results in the activation of related counterintuitive properties. We also did not find a sharp boundary between INT and MCI concepts as some intuitive ideas were rated as less expected than some counterintuitive ideas. The following intuitive concepts were rated by our participants to be more surprising than the counterintuitive concepts given below (mean expectedness ratings are shown in parenthesis besides each statement).

More Surprising Intuitive Statements

Mental being that understands others can also talk to others (mean expectedness: 0.73)
Physical contact is required for launching or changing the direction of movement of all solid objects (0.73)
Solid objects move as connected wholes (0.73)
Animals are self-propelled (0.97)
Solid objects that requires physical contact for launching or changing the direction of movement also moves as connected wholes (1.09)
Animals take actions to satisfy their goals (1.36)
Mental beings have emotions (1.38)
Mental beings have self-awareness (1.39)
Living things produce offspring that are similar to themselves (1.39)
Mental being that eats also has strong beliefs (1.40)

Less Surprising Counterintuitive Statements

Animal that can talk can also understand English (2.11)
A solid object that has processes to satisfy its nourishment needs is also composed of natural substances (1.83)

A solid object that produces offspring that are similar to it also grows (1.83)
An animal that has strong beliefs also has self-awareness (1.74).
An animal that talks also has self-awareness (1.50)

4.3.2.5 No Distinction between Concepts Learned through "Maturationally Natural" and "Practiced Natural" Processes

A problem with the traditional content-based accounts of the minimal counterintuitiveness effect is that by focusing on a problem in isolation, they lack the number of constraints on the mind that are needed to eliminate such parameters. Celebrated cognitive scientist Allan Newell argued in his book "A Unified Theory of Cognition" that "mind is shaped by multiple constraints. It requires a unified theory of cognition to bring these constraints to bear." The minimal counterintuitiveness effect, for instance, cannot be considered in isolation from the well-studied cognitive phenomena such as concept learning, conceptual combinations, and discourse processing. A parsimonious theory of the MC-effect should start from a unified theory of cognition to develop a model that takes advantage of what we already know about how people understand narratives and learn new concepts from them to build a model that does not require ad-hoc parameters (such as where is the boundary between MCI and MXCI concepts. Proponents of the content-based view have placed the MXCI/MCI boundary at different points. Atran (2004) considered concepts that violate 2 category expectations as MXCI while Barrett (2008) labeled them as MCI). This is precisely what the context-based model of attempts to do. The context-based model for instance defines MCI and MXCI concepts not in terms of the number of category expectations they violate but whether or not people can justify them given various contextual variables (e.g., their motivation level, the time available, and cognitive load placed on them).

By building on theoretically sound psychological models, the context-based model not only explains the minimal counterintuitiveness effect in a parsimonious fashsion but it also explains a broad range of findings. One of the most robust findings in experimental psychology has been the so called *distinctiveness effect* which indicates that an item, that stands out as compared to other items in its context, is more likely to be remembered than those other items (Hunt & Worthen, 2006). For over a century, experimental psychologists working with a variety of stimuli have found support for this effect (von Restorff, 1933). Thus unexpected events and entities in a story are recalled better than expected events and entities (Kintsch, 1998), bizarre images are recalled better than ordinary images (McDaniel, Einstein, DeLosh, & May, 1995), unexpected words in a list of words are recalled better than expected words (Fabiani & Donchin, 1995), orthographically distinct words are recalled better than ordinary words, as are typographically distinct words (Hunt & Worthen, 2006).

Semantic memory models such as spreading activation make no distinction between memories formed through "maturationally natural" and "practiced natural" processes and therefore concepts that violate expectations of practiced cognition should also be more memorable than those concepts that do not. Thus the context-based view predicts that those deviations from cultural schemata and scripts that can easily be made sense of by a reader in a given context will be better remembered than information that is typically part of a schema/script. This has been a robust finding in the literature on memory for schema/script-congruent and schema/script incongruent items (Bower, Black, & Turner, 1979). Psychologist Denise Davidson and her colleagues, for instance, found that counterintuitive actions such as "the flowers singing" and "the dog reading a book" and a horse telling people "to get into the wrong line" (what the authors labeled as "implausible actions") were better recalled than intuitively expected script-congruent actions (labeled by authors as "plausible actions") such as "a girl eating pizza", "a boy looking at a carrot", and "a man comparing cheeses" (Davidson, Larson, Luo, & Burden, 2000).

4.3.2.6 Counterintuitive Concepts Should Take Longer to Process Than Intuitive Concepts

The context-based model predicts that, on average, readers should spend more time to process counterintuitive concepts than they do in processing intuitive concepts. This is because counterintuitive concepts trigger cognitively taxing process of justification creation while intuitive concepts do not. When my colleague Professor Mary Harmon-Vukic and I measured the time that it took subjects to read intuitive and counterintuitive concepts, we found that subjects spent more time processing counterintuitive concepts than they did processing intuitive concepts (Harmon-Vukic & Upal 2012; Harmon-Vukic, Upal, & Sheehan, 2011).

4.3.2.7 Counterintuitive Concepts Should Become Less-So Overtime

The context-based model suggests that the same concept may appear more unexpected in context *A* than in context *B* and that the same concept may be more memorable in one context and less memorable in another context. Since knowledge structures in people's memories change over time, the same concept may be more counterintuitive for a person at a time t_1 than at a time t_2. A one-time exposure to an idea, however, does not guarantee that the idea will not seem counterintuitive in the future. In order for an idea to lose its memorability advantages, the knowledge in long term memory that generated the expectations has to be revised so as to make the counterintuitive idea as the new expected and the old idea as the new unexpected (and therefore the new counterintuitive). Since knowledge structures in memory are richly connected with each other, akin to Velcro loops (Heath & Heath, 2007), revising them requires significant cognitive resources to untangle old loops and establish new connections. Thus it is not surprising that people are very conservative when it comes to revising their beliefs.

People's expectations guide what they see, leading them to sometimes miss the unexpected objects and events. When the evidence of expectation violations is too overwhelming to ignore, they prefer to generate elaborations that allow them to preserve as much of their old beliefs as possible. Even though observing a single instance of a counterintuitive object or event can (at least in principle) trigger belief change, this does not happen very often. For instance, upon seeing an ostrich for the first time, one may no longer be surprised when one hears of, "a healthy adult bird that cannot fly" assuming one can create a justification that an ostrich is still a bird because it has feathers but is not able to fly because it is too heavy. Creation of justifications in response to seeing an unexpected object or event does not automatically lead to generation of different expectations in a similar future context. One may for instance assume that the expectation violation only happens in an overly restricted context, for instance, assume that ostriches do not fly on Tuesdays between 9 and10 am or that the ostrich under observation is a mutant bird. Seeing a healthy adult ostrich at a different time in the future may still lead to the expectation that it will fly. It may take prolonged exposure to numerous observations of unexpected and significant cognitive effort for someone to revise enough knowledge structures in their long term memory to generate new expectations. However, once all the relevant memory structures have been revised and the "old unexpected" becomes the "new expected," the once minimally counterintuitive idea are no longer so. Thus the context-based model predicts that minimally counterintuitive ideas should lose their memorability advantages over time as a person is repeatedly exposed to them.

4.4 Chapter Summary

Since my original presentation of the context-based model in 2005, a growing number of studies (Gonce et al., 2006; Slone et al., 2007; Tweney et al., 2007; Sheehy-Skeffington, 2008: Harmon-Vukic et al., 2011; Attia-krieger, 2014) have found that it better accounts for studies that investigate memory for various types of concepts (see the excellent review by Russell & Gobet, 2013). Since the context-based model does not support differential processing for mental knowledge acquired through intuitive and doctrinal modes of thinking, it predicts that violations of online intuitive cognition concepts should not have a privileged status, at least when it comes to memorability. Thus, ideas that involve violating expectations generated by offline learned concepts such as cultural schemas and religious doctrine should also be better remembered than ideas that do not violate such expectations. The context-based view emphasizes the role played by the knowledge that an individual possesses when processing a concept in making a concept a concept minimally counterintuitive. This means that a concept that is minimally counterintuitive for one person may not be minimally counterintuitive for another person whose mental knowledge differs from that of the first person[21]. If counterintuitiveness is not the property of the concept alone, then a concept can only appear minimally counterintuitive to a population if individuals within the population share beliefs that are relevant to understanding the concept i.e., if the concept violates the expectations raised by those shared beliefs and if the expectation violation can be justified using those shared beliefs. I will refer to such ideas as *socially counterintuitive* and argue that ideas need to be socially counterintuitive to spread in a population and become culturally embedded.

5 Social Counterintuiveness

- More than two-thirds of British (68%) believe that "human beings evolved from less advanced life forms over millions of years". This belief is shared by a majority (61%) of Canadians and only a minority (35%) of Americans.[22]
- Eight out of ten adult Americans believe in God. Large majorities also believe in miracles (75%), heaven (73%), angels (71%), that Jesus is God or the Son of God (71%), the resurrection of Jesus (70%), the survival of the soul, after death (68%), hell (62%), the Virgin birth (Jesus born of Mary (61%) and the devil (59%).[23]
- A strong minority (47%) of Americans believe in "Darwin's theory of evolution" while almost equally strong minority (40%) believe in "creationism"
- Significant minorities of Americans believe in ghosts (44%), UFOs 36%, witches (31%), astrology (31%), and reincarnation (24%).

Most cultural scientists believe that shared beliefs form an important aspect of culture. Shared beliefs of a group of people have been studied as ideology, worldview, and belief systems. As Bar-Tal (2000) argues, such shared beliefs play a critical part in formation and maintenance of social groups.

> ... members of all human groups share beliefs. Sharing beliefs is an integral part of group membership. Individuals, as members of different groups, hold shared beliefs in their cognitive repertoire. Some of the shared beliefs serve as a basis for group formation, provide meaning to their group membership, and direct or justify many group actions. (Bat-Tal, 2000: xi)

At least some of people's shared beliefs are theory-like. Note that theory in this context does not mean a complete, organized, scientific, coherent, and consistent system of knowledge but rather fragments of rules and causal-knowledge that can be used to generate expectations about objects and events that the group encounters in its environment. There has been a long running debate among social scientists about the degree to which shared beliefs such as ideologies are constrained to form coherent and consistent belief systems. The "end of ideology" camp argues that ideological labels such as 'conservative' and 'liberal' have little predictive power in terms of their ability to predict political views and behavior of ordinary Americans (as opposed to those of political activists). The resurgence of ideology camp argues that:

> it does not follow that when citizens struggle to articulate a sophisticated, coherent, ideology, they must be incapable of using ideology with either sophistication or coherence. Very few speakers can state precisely the grammatical and syntactical rules they obey when speaking their native languages, and yet they use language adeptly (albeit imperfectly). (Jost, 2006: 657)

This debate on ideology has limited relevance to the point being argued here, namely, that members of a group share knowledge that allows them to generate expectations that they share with other members of their group. This is because a logically coherent and consistent scientific theory is not a requirement for generating expectations. There is an emerging consensus in cognitive sciences that people use their world knowledge to generate expectations about their environment (Schank, 1999). Furthermore, even the end-of-ideology camp admits that holders perceive their shared belief system to be logically coherent (Converse, 1964: 5). This may be because people prefer to believe new information that is logically consistent with their existing beliefs and when people detect inconsistencies in their beliefs, they try to resolve them (Festinger, Riecken, & Schachter, 1964). While this does not guarantee that the knowledge held by individuals is internally coherent and consistent, it does means that it is not an arbitrary collection of random ideas. Developmental and cognitive psychologists have accumulated considerable evidence in support of the "theory-based" view of concept acquisition which suggests that people organize their knowledge as theories and that such knowledge impacts the way people perceive and make sense of their world.

I define an idea as *minimally socially counterintuitive* (*MSI*) for a population if it violates a sufficiently small number of expectations generated by beliefs shared by that population such that these violations can be justified given the group's shared beliefs (Upal 2011a). Thus the notion of a person remembering details of her past lives may be minimally counterintuitive to a Western population that may have a passing familiarity with the idea of reincarnation but not to a Hindu population that has the belief in reincarnation intricately woven into the fabric of its shared beliefs. Minimally counterintuitive social ideas have a memorability advantage over socially intuitive ideas that conform to the expectations generated by the group's shared cultural beliefs and maximally socially counterintuitive ideas that violate a large number of cultural expectations. Thus the notion of a person who remembers a past life as Queen Mary I (1516-1558) would have a memorability advantage in the British population. This is because British people, by and large, do not expect their fellow citizens to remember their past lives but they can use their passing knowledge of reincarnation to make sense of such claims. However, it will not enjoy memorability advantages due to counterintuitiveness in an Indian Hindu population where such ideas are socially intuitive or in a group of 10-year old Saudi madrassa students who have never heard of reincarnation or about Bloody Mary and thus the idea of past-life-as-Mary-remembering-human is maximally counterintuitive to them.

In most situations, MSI ideas are also MCI by definition, because in order to violate shared expectations of a group of people, an idea must violate expectations of individual members of that group. Thus MSI ideas have the transmission advantages that have been ascribed to MCI ideas (Barrett, 2008). However, it is not clear whether MCI ideas that are not MSI would have the transmission advantages that would allow them to become widespread in a population. An idea that is counterintuitive to a few members of a group and not to a large number of group members would only have memorability advantages in that small subgroup. Thus the notion of social counterintuitiveness is more crucial to explaining cultural success of surprising ideas than the traditional notion of individual counterintuitiveness.

While we like to think that we know what most members of our group believe, our perceptions are not always accurate as illustrated by the "emperor is naked" phenomena (Centola, Willer, & Macy, 2005). In situations, where most individuals have accurate perceptions of other people's beliefs, MSI ideas may have additional transmission advantages due to their perceived interestingness to the listener. If John believes that Kate would find idea *X* surprising and therefore more interesting then John may be more likely to tell Kate about *X*.

5.1 Ratcheting-up of Social Counterintuitiveness

Some of my earliest childhood memories are those of Ammiji, my maternal grandmother, holding my hand to help me step over the sewage streams on our way to my cousin Saeed's house two blocks down the street. She was a diminutive but loving woman who helped us read Quran and told us stories as we gathered around her bed. Given her loving nature, it was really surprising (and frightening to us kids) how she would completely forget about us, her daughter, and her husband (my Abbaji) whenever she suffered a schizophrenic episode. She would think that she was a young unmarried woman living in her village in British India and would leave our house to return to her parent's house. Since most of our neighbours in Rabwah knew about her condition, someone would see her wandering around and alert us. In the summer of 1979, however, when she was visiting my aunt in the large urban metropolis of Lahore, she left the house and never returned. Abbaji (my maternal grandfather) wrote to Hazoor[24]–Ahmad's grandson and the reigning Ahmadi Caliph–to pray that Ammiji be found safely. Unfortunately, Hazoor's prayers didn't work and we found Ammiji's body three days later. Abbaji brought her to Rabwah so that she could be buried in the Bahishti Maqbara-the Ahmadiyya heavenly cemetery.

The original heavenly cemetery was established in Qadian by Mirza Ghulam Ahmad, the founder of Ahmadiyya Jama'at, shortly before his death. In 1905, Ahmad published a book titled "Risala Alwasiyyat" (i.e., The Will) in which he said that:

> I was shown a grave which was brighter than silver and all its clay appeared to be silver and it was said to me: This is your grave. I was shown a place that was named Bahishti Maqbarah, and it was conveyed to me that it contained the graves of such selected members of the community who are destined for heaven...
>
> And because very great tidings have been given to me about this graveyard and because God did not only say that this graveyard is heavenly, but also said that every kind of blessing has been descended on this graveyard, and there is no blessing which is not shared by those who

> are buried in this graveyard, God has inclined my heart through His revelation towards the idea that for the burial in the graveyard some conditions should be prescribed and only those would be admissible who, because of their truthfulness and their perfect righteousness, comply with them...
>
> The first condition is that whoever desires to be buried in this graveyard should contribute towards the expenses of its maintenance according to his/her capacity. Such contributions are demanded only from people who desire to be buried herein, and not from others... there should be an Anjuman which should be responsible to spend the funds from such income... from among the Jama'at only those will be buried in this graveyard who make a testamentary disposition that one tenth of his/her entire property shall, under the directions of the Movement, be devoted to the propagation of Islam. (Ahmad, 2005: 22-26)

Since Ammiji had been an active member of the *wasiyyat* system throughout her adult life, having paid off 1/10th of her property and paying her monthly dues, we were shocked when she was denied a burial in the heavenly cemetery. When we asked why, Abbaji's letter to Hazoor was thrown in our face as evidence that Ammiji was not mentally sound. This according to Jama'at officials disqualified her from burial in the heavenly cemetery. We had to live with the indignity of burying her in Rabwah's Commoners Cemetery (*Qabrastan-e-aam*). How did the small tract of hilly land in Rabwah, that bore no resemblance to the green tract of land in Qadian that Ahmad saw in his 1905 divine vision and where he lay buried, come to be considered a heavenly cemetery by Ahmadis?

Ahmad's original idea of a heavenly cemetery in Qadian had been counterintuitive enough for his Ahmadi followers. Even forgetting about the monetary component, simply the notion that people could find out within their lifetime whether they were going to heaven or hell was sacrilegious to most Indian Sunni Muslims. Asking for money for it only made it worse. It was akin to claiming that God was selling pieces of heaven for morsels of gold. A small number of Ahmadis, who were able to justify this counterintuitiveness in part by convincing themselves that the proximity of their burial site to the Promised Messiah's blessed grave may be spiritually beneficial, took part in the scheme and became *moosies*[25]. Even these Ahmadis would have had trouble justifying the claim that a lot of barren land hundreds of miles away in the middle of nowhere in backwaters of the Jhang District inhabited by *jaanglis* (literally savages, the term people of Rabwah use to refer to the Jhang dialect of Punjabi language and its speakers) was also heavenly. That idea probably would have been maximally counterintuitive to most Qadiani Ahmadis had they heard it in 1905. This begs the question of how did such an idea come to be accepted by so many Ahmadis in the 1950s? What had changed to make the idea of a Bahishti Maqbara in Rabwah justifiable to so many Ahmadis who would not have accepted it in 1905?

In order to answer such questions we need to understand that similar to the case with individual counterintuitiveness, socially counterintuitive ideas can also become socially intuitive over time even though the process is more difficult because it involves changes in shared beliefs of a large number of people. As advocates of social change would attest, getting a new idea to become widely accepted by a population is a long and painstakingly slow process that requires years of effort by dedicated individuals. This is because, similar to individual ideas in individual memories, shared cultural ideas are deeply intertwined with each other. Once the well knit-fabric of cultural beliefs is ripped up by a socially counterintuitive idea, a number of formerly tied threads become ripped open. All of these threads have to be re-stitched together in new and innovative ways to fully mend the cultural fabric. Similar to individual beliefs, however, once culturally counterintuitive ideas become seamlessly embedded into a group's cultural scaffolding, they are no longer perceived as culturally counterintuitive by members of that group.

Once a socially counterintuitive idea becomes socially intuitive in a group, it loses its memorability advantages in that group. This resolves another paradox that critics of cognitive science of religion have often pointed out, namely, that while the counterintuitive beliefs such as religious belief in gods, and ghosts and popular culture beliefs about Draculas, vampires, Vulcans, djinns, chupacabras, and leprechauns are counterintuitive in the traditional cognitive science of religion sense, they do not appear to be counterintuitive to the people whose informational worlds are inhabited by such creatures. Theists from a variety of theistic traditions, for instance, routinely point out that they see God in everything such as people's eyes, flower petals, grass blades, running streams, stars, and singing birds and that the concept of God appears no more counterintuitive to them than air, energy, and kinetic potential. The anthropologist Maurice Bloch poses essentially the same question when he asks: "what if a statement seems to us apparently odd, but not so to the people concerned?" After having lived in a culture and being thoroughly immersed in it, even a visiting anthropologist may stop viewing the beliefs that used to appear counterintuitive as counterintuitive. He writes:

> the first time one hears such a story... one treats it as odd, but then, subsequently, one hears this story so often that, every time it comes up, it requires less and less critical attention; in the end, it becomes exactly like an unexamined intuitive belief, in that its hearsay aspect has been eroded. When information has become as familiar as this, it is no different from being told that if you eat too many unripe fruit you will get diarrhea – something you may well not have experienced yourself, and for which you have no intuitive connection between cause and effect, but which seems so sensible... (Bloch, 2005: 106)

He further writes:

> The basis of the problem is that a stance which identifies the counterintuitive with a priori characteristics, that can be inferred from the nature of the entities proposed, ignores the role of communicative practices in real situations... such an approach ignores the everchanging evolution of attitudes of people towards these entities through time and in different circumstances.

> Thus a representation which a particular person might understand as counterintuitive when they first came across it, out of the blue, so to speak, clearly does not have the same cognitive significance as it does when it has become totally familiar, and has been held as valid by oneself and everybody else around for as long as anyone can remember. (Bloch, 2005: 114)

The answer I believe lies in acknowledging the criticism that minimally counterintuitive ideas do indeed lose their privileged status and do not have any memorability advantages once they become embedded as part of a culture. However, this does not mean that further cultural innovation stops. New ideas continue to be created and communicated to others and those ideas that have transmission advantages spread faster than those that do not. In order to have memorability advantages due to counterintuitiveness however, new ideas must violate people's expectations in the new context and not the old context, which is no longer relevant because people's expectations are driven by what they currently believe and not what they (or their ancestors) used to believe. This means, for instance, that once a minimally socially counterintuitive idea such as the idea of "a being who can see everyone" becomes widely culturally accepted, it loses its memorability advantages because it no longer violates people's expectations. In order for a concept to achieve memorability advantages and spread in the new cultural context, an idea has to be seen as counterintuitive in the new context. One way to do that is to build on the counterintuitiveness of the old idea. For instance, the concept of "a being who can see *and hear* everyone" may be seen as minimally counterintuitive in the new context. Note, however, that the new idea may have been maximally socially counterintuitive in the original context. In light of the context-based model, one should not be surprised to see maximally counterintuitive concepts to form a significant part of religious beliefs. Indeed, it would be surprising if they did not!

The ratcheting-up of counterintuitiveness not only explains how seemingly maximally counterintuitive concepts, such as the Judeo-Christian-Islamic God concept, come to be widely distributed but it also predicts a continuous transmission advantage for unorthodox ideas that violate cultural expectations over traditional ideas that do not. This explains the continuing evolution of cultural beliefs among groups ranging from post-modern artists to new religious movements. Cultural historians often resort to using the analogy with waves on a beach to explain the waves of innovation that seem to continually change the landscape of culture. Art historians for instance see the last few centuries of Western Art history as waves of impressionism, expressionism, Fauvism, Cubism, Dadaism, and surrealism etc. Any two historically contiguous waves have an interesting paradoxical relationship with each other. The new trend is both defined in opposition to the old one and also as a continuation and improvement of the old trend. At the core of each trend is a minimally counterintuitive idea that is advocated by a group of innovators and becomes widespread because it is culturally counterintuitive for the population of interest. However, once it becomes widely accepted and integrated into the cultural beliefs of the group of individuals, it loses its memorability advantages making room for yet another wave of innovation to build atop this new informational landscape.

To summarize, the notion of social counterintuitiveness suggests that unorthodox ideas that violate shared expectations of a group of people should be remembered better than other types of ideas and that these memorability advantages should give them an edge in competition with conventional ideas. The social counterintuitiveness model outlined here predicts a dynamically changing face of cultural information where new trends and innovations are constantly being shaped and reshaped. This is certainly what social psychologists working to stem the flow of rumors have found. They have found that rumors are similar to a dynamically evolving flame that is constantly changing direction and building up in certain directions.

5.2 Splits in Religious Movement

Scholars of religion in general, and new religious movement (NRM) scholars in particular, have also documented such trends of innovation in the realm of religious beliefs. The NRM scholars argue that splitting of an NRM from an existing movement often follows the introduction of an innovation into the doctrinal beliefs of the existing movement. Sociologists of Religion, Bill Bainbridge and Rodney Stark, provide a number of examples of new religious movement leaders who created the fundamental doctrines of new religious movements by modifying the beliefs of an existing NRM (Bainbridge & Stark, 1979). Indeed they argue that tracing the history of such deviations, labeled "cultural genetics," may be a useful way to study NRMs. Bainbridge counts over half a dozen movements that split from Dianetics and the Church of Scientology in the short period of 20 years from 1952 to 1972 (Bainbridge, 1985). As we will see in Chapter 8, Ahmadiyya Jama'at's doctrine went through numerous waves of ratcheting-up of social counterintuitiveness. These include beliefs about Ahmad's prophethood, beliefs about Jesus' death, and beliefs about a heavenly cemetery.

Unlike the content-based view which suggests that counterintuitiveness of an idea can be determined simply by considering the contents of the idea in question, the context-based model suggests that the context in which an idea is processed by a social group must be taken into account. The context consists of shared perceptions of members of the group about the concept in question. Thus in order to judge whether Ahmad's ideas were surprising for members of his target audience of North Indian Sunni Muslims or not, we must carry out a deep historical dive to understand shared perceptions of nineteenth century North Indian Muslims. We will pay particular attention to those shared beliefs which are needed to judge intuitiveness (or counterintuitiveness) of Mirza Ghulam Ahmad's claims. We will also be cognisant of social psychological forces at play, in particular, social identity beliefs of North Indian Sunni Muslims to help us understand how counterintuitive concepts come to be integrated into social identity perceptions of a group through painstaking efforts of social identity entrepreneurs such as Mirza Ghulam Ahmad. I hope that this case study also illustrates how I view the cognitive science of religion and sociocultural and historical study of religion to progress in lockstep with each other.

6 Shared Beliefs of Northwestern Indian Muslims

Understanding shared beliefs of a group is key to determining intuitiveness or counterintuitiveness of an idea and thus any transmission advantages its counterintuitiveness may confer on it. In order to understand how Mirza Ghulam Ahmad's ideas were perceived by his primary target audience consisting of late nineteenth century North Indian Muslim elite, we first need to understand their shared beliefs about themselves, about Christian missionaries, and their expectations about imminent arrival of a saviour. This would be really easy to do if modern opinion polls had been regularly conducted among the nineteenth century North Indian Muslims! Since, they were not, we have to rely on writings of the principal actors that are available to us and conduct a historical analysis. Such a deep dive into history may be difficult to follow for some of the readers but according to the context-based view understanding the nitty gritty details of people's mental representations is crucial to being able to explain transmission advantages that some ideas enjoy over others. To those cognitive science readers who wish to skip a detailed historical account, I suggest skipping to Page 91 to read Chapter Summary for Chapter 6 and then to Page 119 to see Chapter Summary for Chapter 7.

6.1 Northwestern Indian Muslim Perceptions of Themselves

The Mughal empire that had dominated Northern India for centuries went into a sharp decline following Aurangzeb's death in 1707 for a variety of reasons not the least of which was infighting among the royal family for succession. With the top layer of this "layered sovereignty" unravelling, various regional leaders, including many Mughal appointed governors, asserted their authority as sovereigns. The most dominant states to emerge from the decline of the Mughal power in North India included the Maratha state and the Sikh state. Marathas, who held a vast territory in Western and Central India, captured Delhi in 1784. Various Sikh principalities to the Northwest of Delhi were consolidated into a state by Raja Ranjit Singh (1780-1839).

In the eighteenth century, Northwestern India contained a significant proportion of Muslims (10-20% of the population by various estimates). This was especially true of the major cities of Lahore and Delhi half of whose population was Muslim. Eighteenth century, Indian Muslims were a very diverse population. Ethnically, they consisted of recent migrants from Central Asia, Afghanistan, Iran, and Arabia and local Indian converts hailing from upper and lower castes and of various Indian ethnic groups such as Punjabis, Sindhis, Kashmiris etc. Religiously, they were divided into Shia (Ithna asharis, Ismailis, Dawoodi Bohras), Sunnis (mostly Hanafis), and Sufis (Chishti, Naqshbandi, Qadri, Suharwardi). Socio-culturally, they were divided into upper aristocratic classes who called themselves *ashraf* (meaning respectable) and labeled others as *ajmal* (meaning commoners). According to Encyclopedia of Islam, the term *sharif* (singular of *ashraf*) is an honorific term used throughout the Muslim and Arab world (Campo, 2009). Most of the Indian Muslim ashraf claimed decent from outside India.

1. Sayyids claimed to be descendants of the Prophet Muhammad through his daughter Fatima and her husband (and prophet's paternal cousin Ali ibn Talib). They use Sayyid, Mir or Shah as a title to indicate their ancestry.
2. Mughals claimed a Turanian (i.e., Persian speaking Turkish central Asian) lineage. They use Mirza or Beg as a title to claim their ancestry.
3. Pashtuns claimed an Afghan ancestry. They use Khan as a title.
4. Sheikhs claimed an Arab decent (though not from prophet Muhammad). They use Sheikh as a title to indicate their lineage.

While many ashraf lived in the 'royal cities' of Delhi, Agra, and Lucknow, which had a flourishing ashraf culture, many also lived in small towns (called *qasbahs*) spread all over Northwestern India. The ashraf society was a Persianate society (Hodgson, 1974) because of its use of Farsi as the language of its various cultural expressions. The center of the ashraf culture was the *darbar* (i.e., court) of the King or a lesser noble. The status of ashraf was not a simple matter of blood-lines, it also required a mastery of Persianate culture including the elaborately meticulous *adab* (i.e., etiquettes) system and Farsi and Urdu poetry. The ashraf culture reached its zenith during the reign of Mughal King Shah Jehan (1594-1666) but had a brief resurgence in the early nineteenth century.

The Central Asian Islam, itself an interesting mix of Hanafi Sunnism and Sufism, when brought into the unique multi-ethnic and multi-religious environment of Northern India blossomed into a unique blend of Hanafi Sufism and local Hindu rituals and customs. In this syncretism, a person could achieve a higher spiritual status by devoting oneself fully to seeking perfection in ritualized prayers, fasting, and giving up worldly desires. God was thought to reward those who resist the temptations of the world and spend a lifetime performing pious rituals with various supernatural powers. These included the power of intercession on behalf of others and the power to request miracles to convince the non-believing. A Sufi, or a seeker of knowledge, could progress from Sufi to pir to wali to ghous to qutb to nabi to rasul. According to Metcalf, "the Sufi elder (known as pir, shaykh, or murshid) was an instructor in spiritual disciplines, a guide to the moral way and discipline (tariqa) that led to the inner realization of the Divine, an intercessor for his followers, and a conduit of divine intervention or miracles (karamat) in everyday life" (Metcalfe, 2009: 8). Once a student was deemed to have acquired sufficient knowledge by his murshid, he was awarded a sanad (degree) signed by his master.

Various kings, including the Mughals, patronized Sufis as "inheritors of charisma (baraka) derived through "chains of intercession" (silsila) from the Prophet himself" (Metcalfe, 2009: 8). Kings were not the only ones who sought intercession from Sufis, ordinary people (Muslims as well as Hindus and Sikhs) also flocked to visit their lodgings and offer them nazar (i.e., gifts). Sufis did not lose their special powers with their death. In fact, they got reinforced and "in a reinforced form continued to emanate from his tomb, from things belonging to him and even from his name" (Suvorova, 2004: 11). The Sufi tombs called mazars were visited by people seeking intercession with God. Thus a dead Sufi's belongings and pieces of his tombstone became prized as tawiz (i.e., amulets) and could be sold by his descendants from his former lodgings (Suvorova, 2004: 8). According to Survova (2004), the tawiz eventually transformed into "a piece of paper with a prayer or a verse from the Quran written on it" custom designed by a Sufi to "ward off the effect of evil eye and black magic, some to cure diseases of the body and others would ensure success in life and so on." By the eighteenth century the mazar "had grown from a modest structure of cubic form with a dome into a prayer and ritual complex (dargah), where side by side with the tomb proper there used to be a small mosque with minaret, living accommodation for the pır ('old man'), or the 'successor of a pır' the sajjadanishın (literally 'one sitting on his prayer rug'), cells for dervishes (hujra), halls for gatherings (majalis), for hearing music (sama) and for celebration of the saint's birthday (maulud) and day of demise (urs), a guest house for pilgrims and also a public kitchen, where any visitor or beggar could get food free of charge" (Suvorova, 2004: 17).

The 'wild growth' of what Suvorova (2004) calls the 'cult of sainthood,' happened despite a perpetual tension between official government appointed Qazis, orthodox ulema, and reformist 'moderate' Sufis such as Shaikh Ahmad Sirhindi (1564-1624) and Shah Waliullah (1703-1762). Both Sirhindi and Waliullah had been formally inducted into the Naqshbandi Sufi silsila. While fully understanding the benefits of a Sufi approach to Islam (e.g., in attracting Hindus and Sikhs to Islam), the Sufi reformers were worried about dilution of boundaries separating the Muslim minority from the much larger Hindu majority. Thus they condemned excesses of Sufism while attempting to show that a moderate Sufism is compatible with a Hanafi Islam. Shah Waliullah, because he grew up in a time of declining Mughal power, was even more concerned with unifying various Muslim factions than he was with reforming Sufism. Shah Waliullah argued that Muslim political unity could only be achieved through an ideological reconciliation between various sects and schools of thought. He preached against taqlid i.e., dogmatically adhering to the teachings of any one of the four Sunni traditional Islamic mazhabs or schools of thought, namely, Hanafi, Shafii, Maliki, and Hanbali. Even though he claimed to be a Hanafi, he said that he would have no qualms in accepting a ruling by Maliki on divorce from an absentee husband if he felt that it was more congruent to the teachings of Quran and the Prophet Muhammad.

Utilizing the authority of Quran and authentic Hadith–sayings of the Prophet Muhammd–Muslim ulema should use their powers of ijtehad to accept rulings by mufti of any school of thought or create an entirely new ruling if no existing ruling is in accordance with Quran and Hadith. Empowering and freeing Muslims from the yoke of traditional thinking, he thought would not only allow the Muslim ummah (nation) to achieve unity that had eluded them for centuries but also allow them to purge Hindu influences such as praying on the shrines of dead saints that had crept into Sufi Islam. Little did Shah Waliullah know that just over a century later his anti-taqlid message would allow Mirza Ghulam Ahmad to question the centuries of consensus on Muslim beliefs about Jesus.

A number of scholars (e.g., Robinson, 2000; Sikand, 2005) of South Asian Islam argue that traditionally ulema had considered themselves, the Muslim kings, and the ashraf as guardians of Islamic culture in South Asia. Sikand argues that during the times of Muslim power in India most "Muslim rulers as well as the ashraf ulema associated with the courts had little or no interest in the 'proper' Islamization of the ajlaf" (Sikand, 2005: 53). Robinson (2000) argues that equipping all Muslims with proper Islamic knowledge increasingly concerned Muslims leaders who were worried about sustenance of Islam without political power in India. He credits Shah Waliullah for placing "on the individual Muslim conscience the burden of responsibility for creating an Islamic society and give him or her the knowledge to enable them to do so" (Robinson, 2002: 108). Shah Waliullah argued for the revision of madrassa curriculum with more emphasis on Quran and Hadith and less on logic and philosophy so that all Muslims could become defenders of their faith. Shah Waliullah translated the Quran into Persian to allow non-Arabic speakers to understand it.

After Shah Waliullah's death, his mission was continued by his four sons, Shah Abdul Aziz (1745-1823), Shah Rafiudin (1749-1817), Shah Abdul Qadir (1753-1827), and Shah Abdul Ghani. Raifudin and Qadir translated the Quran into Urdu, which had become the literary language of North India, to allow all literate Muslims to understand their holy book. Shah Waliullah was succeeded as leader of Madrassa Rahimiyya by his eldest son Shah Abdul Aziz who in turn was succeeded by his nephew Shah Ishaq (1782-1846) because Aziz did not have any sons.

While all modern south Asian Islamic movements revere Shah Waliullah and his sons, his legacy splintered into two groups that mainly differed on whether North India had become a Darul-Harab (land of war) or remained a Darul-Islam (land of peace) despite the dominance of non-Muslims powers. Waliullah's grandson (and Ghani's son) Shah Ismail (1779-1831) joined Syed Ahmed Barelvi (1786-1831) who declared India to be a Darul-Harab and urged Indian Muslims to undertake hijra (migration) to Muslims lands and launch a Jihad against their non-Muslim rulers. Ismail and Barelvi migrated to the Northwestern frontier of India and took up arms against the Sikhs in Punjab. Barelvi labeled his movement Tariqa-e-Muhammadi while others labeled them as Wahabis because of the similarity of their worldview to that of Abdul Wahab of Arabia. The Wahabis were considered extremists by the British government and they came to be persecuted both in Arabia and India. The second faction was led by two former Madrassa Rahimiyya students, Syed Nazir Hussain (1805-1902) and Nawab Siddiq Hasan Khan (1832-1890). They argued that India was a still a land of peace because the British had permitted Muslims to freely practice their faith. They called themselves variously as Muwahideen (i.e., unitarians- the term preferred by Khan) and Ahl-e-Hadith (i.e., the followers of Prophet's Words- the term preferred by

Hussain). The Ahl-e-Hadith took Shah Waliullah's anti-taqlid approach further and argued that Muslims were not obliged to follow traditional rulings of any of the four traditional schools of thought and that they should only use Quran and Hadith as their guidance to make independent judgements of ijtehad. While Khan married the queen of Bhopal and moved there, Hussain stayed in Delhi and became the foremost reformist Muslim leader in Delhi. He attracted hundreds of students from all over India and the larger Islamic world. Hussain was considered to be a Muhaddis by his followers who called him Shaikh-ul-kul i.e., the master of everything knowable. One of his prominent students, Muhammad Hussain Batalavi (1840-1920), was so worried about being lumped with Syed Ahmad Barelvi's Wahabis that he lobbied the British officials to stop referring to Ahl-e-Hadith as Wahabis. In 1876, Batalavi started an Urdu magazine called Ishat-us-Sunnah as a mouthpiece of the Ahl-e-Hadith movement.

The non-denominational Ahl-e-Hadith view was very influential among Muslim reformers in the nineteenth century including Sir Syed Ahmad Khan (1817-1898)-son of a Delhi ashraf family who became one of the most prominent Muslim reform leaders of the nineteenth century. Khan received a traditional training in Quran and hadith in Delhi (some under Shah Abdul Aziz). Khan claimed that Ahl-e-hadith had been the original faith of Muslims from the very early days that had become corrupted in the medieval times.

> Mahomedanism was at first for many long years a pure and simple Theism; but in the second century of the Higra, when the ideas of the learned men as to its principle were reduced to writing, it was divided into four churches-Hanafi, Shfai, Malki, and Humbali. For some time it remained optional for Mahomedans to choose and follow any doctrine of any of these four Churches. When, however, Bani Umaiya and Bani Abbas became kings as edict was issued directing all Mahomedans to embrace the whole doctrine of any one church of the above four. Those who disobeyed were punished... There were still, however, many who clung the true faith in its primitive simplicity, but who dared not breathe their opinion except to a trusted few. Their name was then Ahal-i-Hadis i.e., believers in the sayings of the Prophet, who were not bound down by the doctrines of the four churches... In India, during the Mahomedan rule, the Turk and Pathan kings who were of the Hanafi sect were strictly averse to religious toleration, and the same state of affairs prevailed during the sovereignty of the Mogul Emperors... On the establishment of the British rule, however, owing to the English principle of strict religious toleration, the followers of Ahal-i-Hadis again came to the front and preached openly and fearlessly. (Khan, 1872: 11-12)

In a letter Khan claimed that he had been a practicing Ahl-e-Hadith long before his friend Nazir Hussain.

> I am the one who turned Syed Nazir Hussain into a puritan Wahabi. He didn't use to perform *rafa-yadain* [Ahl-e-Hadith practice of repeatedly lifting and dropping of hands] during *namaz* [ritual Islamic prayer] even though he considered it to be the practice of those who were rightly guided. I said, it is sad that you do not practice what you consider to be a virtuous deed because of what others will think of it. He got up to offer the *asar* (afternoon) prayer and started practicing rafa-yadain. (Ikram, 2003: 69-70)

Khan was a prolific writer. His early writings mostly expressed traditional Islamic views. For instance, in his book "Qaul-e-Matin dar ibtal-e-harkat-e-zamin" (firm assertion about the false assertion of the earth's motion), he defended the traditional Ptolemaic view universally accepted by early nineteenth century Indian Muslims (Powell, 1993: 209). He changed his views, however, as he learned more about the scientific basis for modern Western views through his interactions with British officers, his study of the Western literature, and a 1 ½ year-long visit to England in 1869-70. He became a forceful advocate of integration of Western scientific knowledge with Islamic worldview. Politically, he argued that Muslims and British were natural allies in India. Robinson (1988) argues that Khan used the techniques invented by his friends Syed Nazir Hussain and Sidiq Hasan Khan to reconcile Muslims to British rule and western civilisation.

> Like the Ahl-i Hadith he circumvented the medieval law schools and went straight to the Quran and Hadith as guidance for Muslims. The basis of his exegetical principles was that the laws of Creation were the Work of God and the Quran was the Word of God and they just could not be contradictory - and if they seemed to be so it was because man failed to understand them correctly. So, for instance, he explained apparently miraculous events in the Quran, as Christian apologists might have done similar events in the Bible, as metaphors. (Robinson, 1988: 10)

In 1880, Khan published first volume of *tafsir* (i.e., exegesis) of Quran. In it, he argued that Jesus' ascension to heaven violated God's law of creation. He wrote:

> The Quran makes mention of Jesus' death in four places... Firstly in *Sura Aal Imran*, secondly in *Sura Ma'ida*, ... thirdly in *Sura Maryam*... fourthly in *Sura Nisa'*. Jesus was not killed by the Jews, either by stoning or by crucifixion, but he died his natural death, and God raised him in rank and status... From the first three verses it is clear that Jesus died a natural death." (Khan, 1880: 48)

Khan raised funds to open schools and colleges for Muslims throughout India where they would be taught traditional Islamic subjects as well as Western philosophy and science translated into Urdu by his Scientific Society. Khan opened the Mohamadan Anglo-Indian College in Aligarh in 1870 and dreamt of making it the Oxford University of India. Overtime, Khan's Aligarh movement came to be seen as distinct from Ahl-e-Hadith. Traditional Sunni Muslims thought that Sir Syed and his followers were blind followers of the Western thinking and they mocked them as *naturies*. While Khan built on the political ideas of Ahl-e-Hadith (i.e., loyalty for the British government), another group that came out of Ahl-e-Hadith built on the distinct identity pioneered by Ahl-e-Hadith (e.g., through *rafa-yadain*). The *Ahl-e-Quran* argued that since the Quran was the most authentic of all revelations of God, it alone should form the basis of Islamic ideals.

> As is not unusual with such intense ideological sects, the Ahl-i Hadith split. Towards the end of the nineteenth century a group emerged under Maulana Abdullah Chakralavi in Lahore which accused them of placing excessive reliance on Hadith, indeed, turning them into a second form

of revelation Chakralavi asserted that only the Quran could be used as compulsory guidance, the Hadith referring merely to the human condition of the Prophet. The group, of course, came to be called the Ahl-i Quran. They were even more exclusive than the Ahl-i Hadith, not bothering to raise the question of whether they could pray with others, but establishing their own prayer ritual with a series of distinctive practices including kneeling only on one knee. They prayed only in their own mosques, eliminated funeral and id prayers, and prayers and alms offered for the sake of the dead. Their dispute with the Ahl-i Hadith was so bitter that eventually the government had to intervene to protect Chakralavi's life. Up to the middle of this century the Ahl-i Quran were found mainly in the Punjab. (Robinson, 1988: 7-8)

Two other scholars who studied under Madrassa Rahimiyya ulema, left Delhi in 1866 to found a new madrassa in the qasbah of Deoband. Darul-uloom Deoband established by Muhammad Qasim Nanatawi (1833-1880) and Rashid Ahmad Gangohi (1829-1905) became very influential in institutionalizing Waliullah's non-denominational views into the mainstream Hanafi Sunni thought. Robinson argues that part of Deoband's success was because of its missionary inspired model of raising funds from public and refusing any form of governmental assistance.

The organisation, moreover, unlike earlier Muslim organisations which tended to live little longer than their founders, was bureaucratic; many lessons were learned from the example of mission schools and pre-Mutiny Delhi College. Much effort was devoted to proselytisation; debating was part of the Deobandi training while large numbers of books in Arabic and Persian were translated into the vernacular and spread wide by means of the printing press. A typical Deobandi book is Maulana Ashraf Ali Thanvi's still popular Bihishti Zevar, first published in the 1890s, which offers complete guidance to a woman on how to behave as a Muslim. While should anyone be in doubt on a point of Islamic law they had only to write to the Dar ul-Ifta (office of Legal Judgements) at Deoband to receive guidance. (Robinson, 1988: 5)

While both Deobandis and Ahl-e-Hadith claimed to be Shah Waliulah's intellectual descendants, the Ahl-e-Hadith were seen as more 'puritanical', 'sectarian' (Robinson, 1988: 7) and 'embattled' (Metcalfe, 1982) by most Indian Muslims.

They [Ahl-e-Hadith] quite clearly set themselves apart from other Muslims, sporting their own cut of beard and insisting on their own form of prayer. This latter practice sharply divided them from Hanafis. It was, in communal prayer, a highly visible and extremely annoying distinction: they said 'amen' aloud, lifted their hands at the time of bowing, folded their hands above the navel and repeated the fatihah (the opening chapter of the Quran which Muslims use as Christians might use the Lord's prayer) aloud along with the Imam. The presence of Ahl-i Hadith often caused disturbances in late nineteenth century India and Hanafis resorted to banning them from their mosques. This led to a series of judicial disputes in which, eventually, the Privy Council ruled that mosques should be open to Ahli-i Hadith as well as Hanafis. (Robinson, 1988: 7)

Partly because of the tireless efforts of the Ahl-e-Hadith, Deobandis, and Aligarh movement and partly because of the changed environment in North India the efforts to reform Islam resonated much more widely among Indian Muslims than they ever had in the past. The spread of lithographic press throughout Northern India allowed the nineteenth century reformers to reach Muslims in ways that were unthinkable in the eighteenth century and earlier. Suvorova argues that the British official and missionary attitudes may also have played a role.

> Of course, a certain role in the hardening of the Indian reformers' attitude towards the cult of saints was played even by the position of the Englishmen, who saw in the veneration of pirs and their tombs one of the manifestations of native 'barbarity'. If the Englishmen displayed a certain respect, even if merely formal, for the institutions of normative Islam, popular religion evoked staunch hostility on their part. The Sufi shaikhs and pirs were associated by them with Roman Catholic monasticism –to be frank an incorrect analogy –and that is why in the heat of puritan indignation they stigmatized them as 'parasites' and 'deceivers of the people.' (Suvorova, 2004: 27)

This is not to say that the reformist movement did not face any opposition. The opposition was led by Ahmad Riza Khan (Metcalfe, 1982) who "used his Hanafi legal scholarship to justify Islam as it had been handed down - a custom-laden Islam which was closely tied to the Sufi world of the shrines where believers sought the help of saints to intercede from them with God" (Robinson, 1988: 7). The faction led by Khan called themselves muqalideen (i.e., followers of tradition) and they lumped Ahl-e-Hadith, Deobandis, and Sir Syed Ahmad Khan's Aligarh Movement into ghair-muqalideen (i.e., non-followers of tradition) camp. Others called Ahmad Riza Khan and his followers Barelvis. The Barelvis accused their ghair-muqalideen opponents in general, and Ahl-e-Hadith in particular, of being conceited individuals who put their own judgement above that of Imam Abu Hanifa–the eighth century founder of the Hanafi mazhab. In a counterattack they redirected the label of bidah (i.e., innovative deviations from pristine Islam) that Waliullah had employed in his polemics against excessive Sufi practices, against the Ahl-e-Hadith to argue that various changes (e.g., rafa-yadain) introduced by them were the real bidah. Sir Syed Ahmad Khan's Aligarh movement came in for the harshest criticism because they were seen as denying supernatural powers to the Holy Prophet and to God in addition to the disregard for the traditional mazhabs they shared with Ahl-e-Hadith. According to the Barelvis, the Aligarh movement represented the worst excesses of the ghair-muqalideen.

As has been noted by most historians (Metcalfe, 1982; Robinson, 2000), the atmosphere in nineteenth century India allowed for a variety of Muslim responses to thrive ranging from Jihad of Syed Ahmed Barelvi to modernization approach of Sir Syed Ahmad Khan. As we will see in the next chapter, Mirza Ghulam Ahmad, found some room even in this crowded space for yet one more variety. Next, I review the shared beliefs of North Indian ashraf about Jesus, the Mahdi, the British rule, and tactics of Christian missionaries in Northwest India because they provide the critical context for understanding why Mirza Ghulam Ahmad's claims were accepted by some North Indian Muslims.

6.2 Beliefs about Jesus

> In Richard Burton's account of his covert pilgrimage to Mecca and Medina, he describes visiting the devotional area outside the chamber (hujra, by tradition the room of Muhammad's beloved wife Ā'isha) of the Prophet's mosque. The chamber itself, Burton discovered, was kept out of view by an ornate curtain. Still he reports that on the other side of the curtain were arranged the tombs of Muhammad, Abū Bakr, and ʿUmar. Moreover, Burton adds with amazement, next to these tombs was a "spare place for only a single grave, reserved for Isa bin Maryam after his second coming". If Burton (whose travelogues tend towards the incredible) can be trusted, the Prophet's mosque itself was arranged in keeping with the prevalent Islamic teaching that Jesus escaped death on the cross, that instead God raised him body and soul to heaven, and that God will send him back to earth in the end times. (Reynolds, 2000: 237)

Western Christians are often surprised to find out that Islam has a well-articulated persona for Jesus albeit markedly different from a Christian one. Even though Jesus is one of God's one hundred and twenty four thousand prophets that were sent to all tribes and nations, he is a special prophet in Islam. He is so special that by most counts he is mentioned 25 times, 5 times more than Muhammad himself. Quran confirms Jesus's virgin birth and calls him a "word from Allah" (3:45) who was "aided by the holy spirit" (5:110). He is mentioned along with Moses as a major law-bearing prophet who was given a book containing God's message (similar to the Quran). Quran also confirms Jesus's miracles and his ability to speak in his cradle. Muslims are told that the divine message being revealed to Muhammad is a continuation and completion of the message relayed to Adam, Noah, Abraham, Moses, and Jesus (33:7). This message needed to be repeated because Jesus's true message had become distorted over time and a humble man had been turned into the son of God.

> And behold! Allah will say: "O Jesus the son of Mary! Didst thou say unto men, worship me and my mother as gods in derogation of Allah'?" He will say: "Glory to Thee! never could I say what I had no right (to say). Had I said such a thing, thou wouldst indeed have known it. Thou knowest what is in my heart, Thou I know not what is in Thine. For Thou knowest in full all that is hidden.
>
> Never said I to them aught except what Thou didst command me to say, to wit, 'worship Allah, my Lord and your Lord'; and I was a witness over them whilst I dwelt amongst them; when Thou didst take me up Thou wast the Watcher over them, and Thou art a witness to all things. (Quran 5:116-117)

The Quran also defends Jesus against the purported Jewish boasting that they had killed Jesus.

> That they said (in boast), "We killed Christ Jesus the son of Mary, the Messenger of Allah";- but they killed him not, nor crucified him, but so it was made to appear to them, and those who differ therein are full of doubts, with no (certain) knowledge, but only conjecture to follow, for of a surety they killed him not.
>
> Nay, Allah raised him up unto Himself; and Allah is Exalted in Power, Wise. (Quran 4:157-158)

Lawson and Reynolds carry out two excellent surveys of the medieval Muslim position on Jesus' death (or lack thereof). Reynolds quotes Abu Jafar al-Tabari, the father of Quranic exegetists as saying that:

> the mufassirūn are in agreement that Jesus did not die, that instead he ascended to heaven body and soul while someone else died in his place. Their principal disagreement is only whether God cast the image of Jesus on a number of people, from whom the Jews chose one to crucify, or whether God cast the image of Jesus only on one specific person. (Reynolds, 2009: 41)

So while there was some disagreement among classical Muslim exegetes over events leading to Jesus's crucifixion, it was not about whether Jesus escaped crucifixion on not (they were unanimous that he escaped crucifixion), but rather who was crucified in his place. Judas Iscariot was considered by many exegetists to be the one who was hanged. Since, Quran refers to Jesus's death using the word tawaffa (meaning we caused him to die) in verses besides 4:157-158 (I just quoted above), classical exegetes had to interpret them in a way that made sense given their interpretation of 4:157-158. Reynolds explains various strategies used by them:

> Tafsīr Muqātil accepts that tawaffā refers to God causing a human to die, but he insists that the Quran uses it for Jesus only in reference to his death in the end times, after his return to earth... Some interpreters, Tabarī notes, are of the opinion that when the Quran applies tawaffā to Jesus it refers not to death but to sleep... According to a second opinion, however, tawaffā – when it applies to Jesus – is synonymous instead with qabada, "to seize"; that is, with this term the Quran is not referring to Jesus falling asleep before God took him into heaven, but rather to the act of God taking Jesus into heaven, or to the moment when God took hold of Jesus before raising him to heaven. These two views of tawaffā, of course, redound to precisely the same doctrine about Jesus. In both cases the interpreters are eager to prove that the presence of the verb tawaffā can be reconciled with the doctrine that Jesus did not die, that he was taken body and soul into heaven, whence he will return... Still Tabarī also cites a third view, that tawaffā – even in the case of Jesus –can only mean "to make die". Most traditions that reflect this view reconcile it, as Tafsīr Muqātil does, with the doctrine of Jesus' eschatological return. If in sūrat āl ʻImrān (3) 55 tawaffā appears before Jesus' ascension, then this verse must be read with hysteron proteron or taqdīm al-mu'akhkhar. Yet Tabarī also notes that some scholars concede Jesus did indeed die. One tradition to this effect insists that he was dead for three hours (another version has seven hours). In the end, however, Tabarī declares his support for the second view, that tawaffā refers in the Quran to God taking hold of Jesus. He justifies this position by referring to the preponderance of hadīth in support of it, but there are other factors at play here. First, for Tabarī the doctrine of Jesus' eschatological return is beyond any doubt. This leads him to reason, in light of quranic passages which imply that a person can only die once (cf. Q 6.60; 19.33), that Jesus must have been preserved from death. In other words, if Jesus is to return in the eschaton to finish his life and die, then the view that Jesus has already died must perforce be rejected. (Reynolds 2009:247-248)

After reviewing positions of the very first Muslim exegetists (such as Abdullah Ibn Abbas who died in 687 AD), Lawson (2009) concludes that, "even at this early date the Muslim community was in agreement on the event of the crucifixion, but not on who was crucified, except that it was not Jesus" (Lawson, 2009: 47). Reynolds (2009) concludes his survey of traditional Islamic views on Jesus' death by saying that classical interpreters of Quran (and by extension ordinary Muslims) believed that God had avoided Jesus' ignoble death on the cross by raising him to the heavens where he awaits his return at the end of times to lead Islam to victory over Christians, Jews, and other non-Muslims. This is the view that Christian missionaries encountered when they followed colonial militaries around the greater Middle East. The sharp decline in Muslim power around the world convinced many Muslims that time for Jesus' return was imminent. The streak of millennialism running through various Christian missionary movements may have only encouraged this belief among Muslims. Powell (1993) describes the 1833 missionary visit to India by Rev. Joseph Wolff of the Society for Promoting Christianity.

> The aspect of his preaching which distinguished him from other missionaries of his day was his conviction that the second coming of Christ was ordained for the year 1847. His own calculations were based on various Old Testament prophecies, principally in the Book of Daniel, but his millenarianism had also been nurtured by contact with the "Irvingites" in England, including Edward Irving, the founder of the Catholic Apostolic Church, Henry Drummond MP, and Hames Heatley Frer. When Wolff reached Lucknow in 1833 after more than a decade of missionary travels, the imminence of the apocalypse was the main theme of his message. (Powell, 1993: 119-120)

Powell (1993) argues that nineteenth century Indian Muslims ulema were so interested in hearing more about Wolff's prophecies about Jesus' return that they overcame their customary reluctance to engage missionaries in a debate (Powell, 1993: 123).

6.3 Beliefs about the Mahdi

The Arabic word Mahdi means guide. Although the word is never used in the Quran, several ahadith (sayings of the prophet Muhammad recorded over a hundred years after his death) mention it along with the sign of the end times. Of the six ahadith books considered authentic by Sunni Muslims, only three (ibn Majah, Abu Dawud, and Al-Tirmazi) chose to include Muhammad's sayings about Mahdi. The two authoritative of the six authentic books (namely, Sahih Bukhari and Sahih Muslim) do not include any ahadith about Mahdi. According to the ahadith included by ibn Majah, Abu Dawud, and Tirmazi, Mahdi and his father will have the same name as the prophet Muhammad and his father. Mahdi's arrival will be accompanied by the return of Jesus. Jesus will assist the Mahdi in killing Masih ad-Dajjal (the false messiah). The Mahdi will usher peace and justice on earth and establish a moral system. Upon Mahdi and Jesus' natural deaths, the angel Israfeel will blow his trumpet to indicate the end of the world. This will be followed by resurrection of all humanity and Judgement Day.

Shia tradition differs somewhat from its Sunni counterpart. According to Shia beliefs Mahdi is already born but is currently hidden and will reappear at the end of times. According the Encyclopedia of Islam, this tradition started with the death of Ali's third son Muhammad ibn Al-Hanafiyya as:

> ...his followers began insisting that their leader was not dead but rather hiding in a transcendent realm from which he would one day return to fill the world with Justice, they initiated a doctrine that eventually became one of the central tenets of Shiism: The occultation (ghayba) and return (raja) of the Mahdi.
>
> The doctrine of occulation and return was developed even further after the sudden death of Ismail ibn Jaafar (d. 762) who had originally been designated the seventh Imam. When Ismail was replaced by his younger brother, Musa a-Kazim, a small group of Shiis calling themselves the Ismailis refused to accept the new Imam and instead claimed that Ismail was alive and in occulation as the Hidden Imam, another term for the Mahdi. For the majority of Shiis, however, the line of Imams continued through Musa until the 12th Imam, Muhammad ibn al-Hasan (also known as Muhammad al-Mahdi), who himself went into final occultation in 941 CE as the Mahdi. (Campo, 2009: 447)

According to ibn-Khaldun, the fourteenth century Muslim historiographer, the concept of Mahdi seeped into Sunni thought from Shias (Khaldun, 1967). The popularity of the concept in Sunni thought seems to have waxed and waned throughout history. The concept was more popular during times of socio-political turmoil when Muslims felt so frustrated with their current leadership that they longed for a saviour messiah to save them. Throughout Islamic history numerous people claimed to have been Mahdis. "From India to Tunisia, every few generations an individual lifts a banner and claims that he is the Mahdi" concludes historian Michal Rubin Furnish (2005). He further argues that despite the "stereotype that Mahdism is chiefly a Shi'i phenomenon", "most Mahdist movements have sprung from the brows of charismatic Sunni holy men" (Furnish, 2005: 5). These movements aimed to restore Islam to its pristine form as was expected of a reformer Mahdi.

> The Sunni Mahdi, more than anything, "was believed to be a divinely guided and appointed agent for renewal (mujaddid), in contrast with the more illuminationist and incarnationist Shi'a ideas." The Sunni Mahdi does not question the validity of the sharī'a in its current form, only the adherence to it during periods of perceived moral laxity. His function is to "support and restore the Sunna of the community, not to transcend or destroy it." This forms the crux of the Mahdi's ideology. In theory, the Sunni Mahdi would agree with the 'ulama's authority, but not with their permissive and lenient interpretation of Islamic tradition. The Mahdi acts as the renewer of Islam, leading to a restoration of Islam to its previous austerity. (McLellan, 2012: 80)

McLellan argues that the Sunnis do not expect the Mahdi to be "an initiator of a new religious dispensation" or to significantly reinterpret traditional Sunni doctrine (McLellan, 2012: 80-81). They only expect the Mahdi to purge Islam of local cultural influences and other "innovations" to restore it to the way it was in the time of the prophet Muhammad. This argues McLellan is why Sunni Mahdis seem to appear more often than not in "the borderlands of Islam" (McLellan, 2012: 81).

Even though Arab Muslims conquered Sind within eight decades of the prophet Muhammad's death, they did not expand their southern kingdoms north or eastwards towards the rest of India. That had to wait for another three hundred years until the crowning of Mahmud Ghaznavi in Afghanistan. He led a wave of raids on temples throughout Northern India but failed to establish a permanent presence. The Delhi sultanate founded by Qutubudin Aibak in 1192 started Afghan/Turkmen Muslim rule over Northern India which lasted until the Marathas gained power over Delhi to be followed by the British in 1803. The declining power of Muslims empires (Mughals in India, Ottomans in Middle East, North Africa and Europe) and the rising power of Christian Europe was deeply worrying for many Indian Muslims especially the *ashraf*. Many thought that this would be an appropriate time for the arrival of the Mahdi to reinvigorate Islam. As Indian Historian Charles Allen recounts:

> ...there was talk of prophecies being fulfilled and of the approach of the end of the days... both Sunnis and Shias shared the belief that at the end of the days a messiah-figure known as the Mahdi, or the 'expected one', would come to the rescue of Islam. He would return to Mecca at the head of all the forces of righteousness to take on the forces of evil in one final apocalyptic battle, after which he and the lesser prophet Jesus would proceed to Jerusalem to kill the devil... in January 1810, when a Muslim named Abdul Rahman proclaimed himself the Imam-Mahdi, collected a band of followers of the Bohra sect of Sunnis and seized the fort of Mandvi in Eastern Surat. The insurgents had then marched on the nearest town, calling on all Hindus to embrace Islam or be killed. The British agent at Surat had been sent a demand to convert, and he had responded by summoning troops from Bombay. Four companies of infantry and two troops of cavalry were landed on 19 January and a one-sided encounter followed in which the aspiring Imam-Mahdi and some two hindered insurgents were killed, after which the uprising fizzled out." (Allen, 2006: 74)

Allen (2006) devotes most of his book to the more successful Mahdiship of Syed Ahmed of Bareli. Syed Ahmed Barelvi not only managed to collect a much larger group of followers including Shah Waliullah's grandson Shah Ismail but also succeeded in establishing a Mahdi-kingdom in the mountainous northwestern frontier of India. He appointed himself as the *Amir-ul-Mumineen* (leader of the faithful) of his Kingdom and ruled a large area which at one point included the city of Peshawar. Allen argues that the reason for Barelvi's success had been the "well established predisposition among all sections of the Muslim community in India to respond to the call of the true Imam-Mahdi in a time of religious crisis, and this now became an established part of Barelvi's Wahabi platform in India: the belief that the end of the days was drawing nigh and with it the imminent return of the Hidden Imam-Mahdi." Canadian Anglican missionary Revd. Worthington Jukes adds:

> Much about this time there was much talk among Muhammadans about the "Imam Mahdy", I wrote a paper on the subject, entitled "Imam Mahdy, and Dajjal, the Muhammadan Antichrist", who was expected to bring the world to a close at the end of the 13th Cent. of the Muhammadan era, and that it was to be synchronous with the return of Jesus Christ, which the Muhammadans

> professed to look forward to. It was published in the October number of the Church Missionary Intelligencer 1883. (Jukes, 1925: 112)

Barelvi and hundreds of his followers died in an 1831 battle with Sikhs who ruled Punjab at the time but it failed to extinguish the Mahdi movement from India. A mere fifty eight years after Barelvi's death, Indian Muslims witnessed the most successful Mahdi movement to originate from British India: The Ahmadiyya Muslim Jama'at. Claiming to be heirs to Barelvi's message, followers of Mirza Ghulam Ahmad would claim that Barelvi had been to Mirza Ghulam Ahmad what John the Baptist had been to Jesus. But before we can turn to their story, we need to better understand the fiercely competitive religious milieu that prevailed in nineteenth century in India in which Christian missionaries were seen to be supported by their compatriot British government workers in their attempts to convert Indian Muslims to Christianity.

6.4 Perception of the British Rule

As discussed earlier, waning Mughal power following Mughal emperor Aurangzeb's death created a vacuum which encouraged various ambitious regional leaders to fight with each other to occupy the Mughal territory. The Indian Muslim ashraf, who were most closely tied to the Mughals in a patronage network suffered greatly from the perpetual fighting that lasted a greater part of the eighteenth century. Delhi, the capital city of Mughals, suffered particularly from the fighting. According to Naim (2001):

> ...the human tragedy of Delhi in the second half of the eighteenth century was indeed immense. The wars between the Turani and Irani factions, the cataclysmic invasion by Nadir Shah, the repeated scourges of the Afghans, the Marathas, the Ruhilas and the Jats—they all took heavy tolls in human lives and also forced much emigration from Delhi and its environs. Then there was the great famine of 1782 in which, according to some estimates, nearly one-third of the rural population of the territory around Delhi starved to death... when the British took Delhi they found that "it had been divided into spheres of control by neighbouring Gujar tribes for purposes of plunder." (Naim, 2001: 8)

Various Sikh principalities to the Northwest of Delhi were consolidated into a state by Raja Ranjit Singh (1780-1839). Ranjit Singh made the old Mughal city of Lahore his capital in 1799. Ranjit Singh considered the territory under his rule to be a holy Sikh state, calling it *Khalsa*, using the name given by Guru Gobind Singh (1666-1708) to the order of initiated Sikhs . According to historian Nadhra Naeem, "Ranjit Singh struck all coins in the name of the Khalsa... He used to call his court the Khalsa *darbar*, his army the Khalsa and himself a humble servant of the Guru" (Naeem, 2010: 296). Ranjit Singh patronized both Sikh and Hindu temples by donating to them generously (G. Singh, 1952).

> The Punjab Akhbar reported on June 13, 1839 that on Sankrant Day, when the Maharaja was gravely ill, he gave away eleven cows with their horns covered with gold, two horses, an elephant, two diamond rings, ten golden and silver images, five golden deer and as many of silver, eleven coral things and two thousand rupees to the Brahmins (Ganda Singh, 1952: 45). Misr Ram Kishan reported on June 24, that he had prepared a golden chair, a bedstead, plates and many other items amounting to twenty one lakhs of rupees to be given as alms (Ibid, 59). A day before his death, the Maharaja also tried to give away the famous Koh-i-noor diamond to the Jagannath Temple in South India dedicated to the worship of Vishnu and Krishna, but Misr Beli Ram declined the Maharaja's orders stating that all assets now belonged to Kharak Singh. As compensation for Koh-i-noor, two armlets with diamonds, worth two lakh rupees, several other jewelry pieces, eight Persian-style top-hats, two elephants with gold howdās and five lakh rupees in cash were given away in sankalāp (charity). (Naeem, 2010: 298)

Ranjit Singh's support of Hindus extended beyond temples and Brahmins. Following the Hindu tradition, he forbade the slaughter of cows by meat-eating Muslims. It appears that at times, Ranjit Singh also forbade the Muslim call to prayer, the azan and "viewed the Muslim clergy with suspicion… as they would promote extremism and rebellion" (Masson, 1842). He also confiscated parts of the jagirs from some of the Muslim ashraf that had been awarded to them by the Mughals. The celebrated Sikh historian Khushwant Singh notes the confiscation of the estates of many Punjabi nawabs by Ranjit Singh's Khalsa darbar.

> The Nawab of Jhang, who had been in arrears for three years… his estate yielding four lacs of rupees a year, was attached to the Durbar. The district of Ucch was likewise taken over. At the same time, on the death of Ramgarhia misaldar, the estates of the misl, which were worth four lacs of rupees a year and included important towns like Qadian and Gobindpur and many powerful fortresses, were attached… (Singh, 1963: 249)

As discussed earlier some of the followers of Shah Waliullah, including his grandson Shah Ismail and Syed Ahmed Barelvi decided to wage a military Jihad against the Sikhs. Others looked to the regional powers to rid them of the repressive rule by the Sikhs and Marathas. For instance, the Muslim nawabs of the state of Malerkotla "accepted British protection in 1809, thereby preserving their territorial integrity and some degree of autonomy" (Bigelow, 2005: 65). These Muslims welcomed the East India Company's intervention in Punjab and supported it in its wars against the Sikh army. They celebrated the defeat of the Sikh army in 1849 and the freedoms of being able to say azan and slaughter cows that came with it.

Contrary to the traditional anti-colonial and anti-orientalists narratives (Said, 1978), recent scholarship (Nizami, 1983; Powell, 1993) has found that the North Indian Muslim ashraf culture (especially around the Mughal darbar in Delhi) went through a revival immediately following the takeover by the East India Company in 1803. After militarily defeating the Marathas, the British allowed the Mughal King, Bahadur Shah Zafar to live in his Red Fort and "maintain a façade of Mughal sovereignty" (Spear, 1969: 88) with the company only acting as his vassal. The "illusions of" Mughal court's "former grandeur were to some extent still fostered by the outlook of the first generation of

British officials in Delhi, most of whom belonged to the school of thought... which held Indian culture and traditions in some respect" (Powell, 1993: 43-44). The company took over the "upkeep of the royal court, the core of the city around the Red Fort palace enclave... to allow the restoration of the royal darbar with considerable pretensions to its former grandeur" (Powell, 1993: 48). According to Naim:

> The Emperor's elephants paraded through the city in festive processions, and ceremonial durbars were regularly held in the Red Fort. It is also true that the people of the pre-Revolt Delhi did not imitate the British in dress, food, and social behavior, and no doubt the etiquette of the royal court was emulated in all élite assemblies in the city, as it was in many similar gatherings all over India... the Emperor regularly took part in the two annual Eid gatherings in the Jama Masjid, and his name was mentioned in the Friday *khutbas* in Delhi as well as elsewhere. His symbolic position as the champion of the Sunni faith also remained important. (Naim, 2001: 12)

The company officers not only took over patronage of Muslim ashraf but to some extent assimilated themselves into the Muslim ashraf culture. Powell discusses how some British officers not only learned Farsi but also acquired court etiquettes (*adab*), attended poetry gatherings called mushairas, and some even wrote Farsi poetry (Powell, 1993: 50-52). As Powell notes:

> ... in Delhi (1806-27), British officials seemed to combine Company service with considerable relish for a 'nawabi' style of comfort and culture. Many of them, including the Residents, Sir David Ochterlony and Sir Charles Metcalfe, had well-born Indian wives or mistresses, and built for their families grandiose town and country mansions in a combination of Mughal and European architectural styles. In this setting they played out roles not unlike those of *umara* in the great days of [Mughal] empire. (Powell, 1993: 51)

Powell describes in detail the case of the company Agent William Fraser who confided in a friend that:

> ...he had never found a 'companion to his taste' among his British colleagues, but companions he certainly had among the ashraf of Delhi. Like Ocherlony, and others who had discovered India before the turn of the [nineteenth] century, Fraser had an Indian wife and several children, whose portraits he commissioned from local artists. His habit of 'consorting with the grey-beards of Delhi' which marked him off from his British colleagues, reflects a number of close friendships with scholarly Muslims... Notables were Shah Abdul Aziz, of the Madrassa-I Rahimiyya, with whom Fraser read Persian, and the poets, Asad Allah Khan Ghalib who, it is said, was 'genuinely attached' to Fraser, and Mufti Sadr al-Din Azurda, who apart from his patronage of literary gatherings, was a key figure in judicial and educational spheres. (Powell, 1993: 52)

The company officers had become so well established as patrons of the ashraf culture that, Ghalib, arguably the father of Modern Urdu poetry, did not find it odd to persistently request the British for a pension for his participation in the darbar activities (Naim, 2001).

The East India company also, at least prior to 1813, "forbade the initiation of any new missionary ventures within its territories in India" (Powell, 1993: 78). Powell says that some of the company directors and officers were 'hostile' to Christian evangelicals because of their reputation as overly enthusiastic unsophisticated simpletons who lacked 'theological complexity' (Powell, 1993: 77-78). Carson points out that the old established 'high church' in Europe looked down at the evangelists as 'fanatical and subversive of the established order' Some company officials wanted to "keep out 'undesirable' Europeans who might disturb the status quo" and "lower Europeans in the eyes of the people" (Carson, 1988: 36). Perhaps not keen to show their newly cultured Muslim ashraf friends, their closeted unsophisticated zealot compatriots, or because of the 'fears of alienating' the natives 'in a land where they were vastly outnumbered' (Carson, 1988: 23-24), the company officers made "strenuous efforts to prevent settlements of any missionaries" in the early part of their rule (Powell, 1993: 79). This policy appeared to have support from Lord Cornwallis, the Commander in Chief of British India and Governor of Bengal Presidency, who is considered to have laid the 'foundation for British rule throughout India' (Dupont, 2001: 483). Rejecting a scheme by Charles Grant to set up company sponsored missions in India, he said that, "he had 'no faith in such schemes', thinking they 'must prove ineffectual'" (Carson, 1993: 45).

After years of Evangelical lobbying in the UK (Carson, 1988), the company decided in 1813 to allow "British missionaries, and from 1833 all Protestant missionaries, to work in India" (Jeyaraj, 2008: 32). Powell (1993) notes two significant changes in the attitude of the British officials in India that happened in the middle of nineteenth century as the British hold on India became more firmly established. Unlike the first generation of East India company officials, the later company British officials were less respectful of Indian traditions and more supportive of a civilizational and Christianisation project for India. These changes argues Powell, "would create a situation in which intrusion into Muslim circles became for the first time a noticeable phenomenon" (Powell, 1993: 79). The great mutiny by some of the company's native soldiers against their British superiors broke out in 1857. This is what Lord Cornwallis had wanted to prevent through his policy of non-interference into religious sensibilities of Indians. According to Carson, he had argued that British officers should pay "a minute attention to the customs and religious prejudices of the sepoys" because "you need not be told how dangerous disaffection in our native troops would be to our existence in this country." (Carson, 1988: 36).

During the mutiny, the Ahl-e-Hadith leaders such as Syed Nazir Hussain and Nawab Sidiq Hasan Khan refused to join other Muslim leaders in issuing fatwas against the British rule. They argued that a religious Jihad could not be waged against the British because they had allowed Muslims to follow all aspects their religion. Instead, both Syed Nazir Hussain and Sir Syed Ahmad Khan actively worked to save the lives of British officials and their families. Syed Nazir Hussain gave refuge to a British woman and hid her in his house till the rebellion was over. Once the mutiny had been crushed, Queen Victoria assumed direct control of the colony and promised to restrain missionary activity and non-interference in religious and cultural customs of Indians. The Queen's declaration of 1 November 1858 was widely circulated in India (Smith, 1923: 728-729).

> It will be our royal will and pleasure that none will be in (anyway) favoured, none molested or disquieted, by the reason of their religious faith of observances, but that all alike enjoy the equal and impartial protection of the law; and we do strictly charge and enjoin all those who may be in authority under us that they abstain from all interference with the religious belief or worship of any of our subjects on pain of our highest displeasure. (East-India-Proclamations, 1908)

This statement was greeted with hope and some skepticism by Muslim ashraf who expected a much worse fate following the uprising of 1857. The missionaries themselves, argues Powell became more cautious in their evangelical activities because they were aware of the "charge that missionary provocation was the primary cause of the uprisings" (Powell, 1993: 283). The British officials, who had started to openly associate with Christian missionaries, became more cautious and adopted a neutral approach in religious disputes.

> In the aftermath of the risings most of the evangelical government servants followed Sir William Muir in avoiding any direct association with controversial missionary publications or with public debates. The Punjab was particularly noted for the evangelical sympathies of its leading officers in this era, but it is significant that two of its Chief Commissioners, John Lawrence and Donald McLeod, in spite of wanting to do 'Christian things in a Christian way', took a firm stand on the principle of 'neutrality'. Herbert Edwardes, a Punjab officer who was censured from Calcutta in 1858 for advocating a display of 'open Christianity' by, among other means, the introduction of the Bible into government schools, was regarded as unwise by most of the evangelical officers in the north-west, who remaining, for the most part, just as committed to missionary objectives as they had been before the risings, were more cautious in their expression of that commitment. It is unlikely, however, that any of these internal changes of emphasis would have been evident to those who were the objects of missionary proselytism. From their perspective of the alim and the pandit the Christian missionary activity in their midst seemed to continue just as intrusively as before the uprisings. (Powell, 1993: 284)

This meant that while the image of British officials as patrons of Muslim ashraf culture continued to suffer, it did not get to the point where most North Indian Muslims perceived them as outright enemies of Islam and in cahoots with the missionaries. Instead, the British officials came to be seen as distinct actors primarily interested in maintaining law and order so that trade could be 'carried on as smoothly as possible' (Carson, 1993: 23). The North Indian Muslims understood that while some of the British officials were personally sympathetic to the cause of Christianisation of India but, wanting to be seen as fair and impartial, were unwilling to openly support the missionaries.

6.5 Perceptions of Christian Missionaries

> Although it's easy to see Christianity and Islam as vast and static forces, they are perpetually in flux. Over time each religion has shaped the other. Religion is dynamic and fluid. The most often overlooked fact of religious revivals... is that they give rise to divisions within the religions themselves. They are about a struggle over who speaks for God-a confrontation that takes place not simply between rival religions, but inside them. (Griswold, 2010: 12)

Eighteenth century witnessed one of the greatest religious revivals in history. The evangelical movement began in the Britain and New England in the 1730s. Evangelicals focused on the role of individual laity in consciously accepting God's gift of salvation as laid out in the Bible without requiring an interpretation by the clergy or the church. Only by deciding to dedicate one's life to Christ- thereby being "reborn"- could one be saved. Great emphasis was also placed on the need to convey this message to those who had not been reborn yet. "Many saw it as their duty to reach new believers, a project known as the Great Commission and rooted in Jesus' parting command to his disciples, "Go therefore and make disciples of all nations, baptizing them in the of the Father and of the Son and of the Holy Spirit" (Matthew 28:19)" (Griswold, 2010: 28). Bebbington (1989) argued that "there are four qualities that have been the special mark of Evangelical religion: *conversionism*, the belief that lives need to changed; *activism*, the expression of the gospel in effort; *Biblicism*, a particular regard for the Bible; and what may be called *crucicentrism*, a stress on sacrifice of Christ on the cross" (Bebbington, 1989: 3).

Perhaps, no other Christian movement besides the Crusades has had as much of an impact on Islam as did the Evangelical movement. This is, in part, due to the character of the movement but also due to the time at which it happened. The eighteenth and the nineteenth centuries marked the heyday of Western colonialism with most of the world having been carved up by European colonial powers. The same European advancements in transportation and communication technologies that made colonialism possible, also allowed British and American evangelicals to reach people around the world. As Griswold convincingly argues, colonial evangelical preachers found Islam as standing in the way of bringing the whole world into the Christian fold:

> "evangelize the world in this generation." These were the words of the Reverend Arthur T. Pierson, a now largely forgotten Yankee evangelical who inspired a worldwide movement. As Pierson put it, "[A]ll should go and go to all." With his urging, the Young Men's Christian Association (YMCA), founded in London in 1844 launched the Student Volunteer Movement. Thousands of young men and women mobilized as missionaries to reach what they believed were the last blank spaces of the map with Gospel... This worldview, with its emphasis on the language of light and darkness, good and evil, flourished in opposition to an enemy. Islam, many evangelicals believed was their most formidable foe. (Griswold, 2010: 29)

India, because of the size of its non-Christian population, was considered a prime target for the evangelization effort. Evangelicals, as early as 1694, had started developing detailed plans for setting up a "government-supported missionary activity" (Carson, 1988: 19). Humphrey Prideaux argued that "English East India Company was declining while the Dutch Company was thriving and attributed this to God's curse on Britain for neglecting the progress of Christianity" (Carson, 1988: 19). The first missionary societies, namely, the Society for the Propagation of Christian Knowledge (SPCK) and the Society for the Propagation of the Gospel in Foreign Parts (SPG), were formed at the turn of the century. Both societies were interested in sending missionaries to company territories in India but they faced a number of obstacles. The biggest problem was finding volunteers willing to risk their lives to live in a far off hostile territory where people did not understand any European language (Carson, 1988). Lack of money to fund travel and salaries was another hurdle. Given the reports from Dutch and Portuguese missions in Goa, Ceylon and Southern India, it was clear that locals were hostile to foreign missionaries. The societies appealed to the company officials in UK for protection and support. The company responded that they welcomed chaplains to minister to the needs of the Europeans in India who were of "tempers and qualifications fit for the undertaking" (Carson, 1988: 28).

The insistence on a proper temperament was meant to exclude the 'enthusiastic' and 'democratic' challengers of the establishment church. Carson quotes Sidney Smith as calling such people 'dangerous' and 'maniacs.'

> ... even for missionary purposes... the utmost discretion is necessary; and if we wish to teach the natives a better religion, we must take care to do it in a manner which will not inspire them with a passion for political change or we shall inevitably lose our disciples altogether... (Carson, 1988: 63).

He argued that Evangelicals were "disgusting and dangerous enough at home" and asked "why are we sending out little detachments of maniacs to spread over the fine regions of the world the most unjust and contemptible opinion of the gospel?" (Carson, 1988: 63). Carson says that the Bishop of Worcester, Samuel Butler, also warned the government that "unless [it] act cautiously, these methodistical proselytizers, by their absurd enthusiasm, will bring about the loss of India." (Carson, 1988: 63). To the extent that prejudices of the high church were shared by decision makers in the British government, such as Lord Cornwallis, their lowly image was a problem for Evangelicals. Evangelicals engaged in an effort to rebrand themselves as respectable subjects of the crown. Carson argues that "a feature of evangelical writings and speeches during the 1790s and 1800s therefore was a concern to emphasize the role of Christianity as a stabilizing force and the best way of maintaining order in society" not just in Britain but also in India (Carson, 1988: 69). Charles Grant argued that "the establishment of Christianity in a country does not necessarily bring it a free political constitution" (Grant, 1813: 96).

Part of this rebranding effort, argues Carson was the formation of new missionary societies, such as the Church Missionary Society (CMS), founded in 1799, and the British and Foreign Bible Society, formed in 1804. These organizations were "deliberately aimed at attracting the aristocracy and episcopacy" to their cause (Carson, 1988: 267). An organized campaign was launched to pressure British Parliament into forcing the East India Company to support missionary work in India. The basic message of the campaign was that (1) Providence has given India to Britain for a higher purpose, (2) that purpose is to bring about spiritual salvation and material improvement in the lives of all Indians, and (3) the only way to help Indians is to convert them to Christianity (Carson, 1988: 256). William Wilberforce of the British and Foreign Bible Society emphasized:

> the importance of obtaining as many as possible of the friends of humanity who may not agree with the us in religious sentiments. All surely will join who do not wish to see such as vast body of our fellow subjects... sunk in the greatest moral and social and domestic barbarism without an effort to raise them on the scale of beings" (Carson, 1988: 288).

A key to selling this message was building an image of poor Indians as being held back by wretched Hindu, Muslim and Sikh ideologies and desperately in need of help. This involved painting 'images of degraded Hindu, and the horrifying practices of sati, hook-swinging and infanticide' (Carson, 1988: 262). Coming on the heels of the anti-slavery campaign, they framed their campaigns as 'moral crusades to ameliorate the plight of Britain's heathen subjects' (Carson, 1988: 271). The evangelicals spread their message through tactics such as, "lobbying key ministers, involving the Established Church, adopting a high moral and religious tone, establishing a nation-wide network" (Carson, 1988: 271).

Evangelicals used various world events and a growing religious awareness in Europe to their advantage. They argued that events such as "the Wilkes agitation, the Gordon riots, the loss of the American colonies and, most dramatic of all, the French Revolution and its aftermath" were "signs of God's displeasure" (Carson, 1988: 255). These events were signs of the end of the world. Carson notes:

> Millenarianism had a close connection with the idea of missionary activity because of the belief that men must labour for souls in preparation for Christ's second Advent: 'And the God of the kingdom shall be preached in all the world for a witness unto all nations: and then shall the end come' [Matt 24.14] (Carson, 1988: 255).

The campaign was so successful that "between April and June 1813 nearly 900 petitions signed by half a million people were presented in the House of Commons in support of the principle that Britain had a duty to propagate Christianity in the areas under her control" (Carson, 1988: 282). On 20 July 1813, the British Parliament voted to include the "pious clause" in the renewed charter of East India Company. It added to the company's mission the objective to:

> promote the interests and happiness of the native inhabitants of the British dominions in India, and such measures ought to be adopted as may tend to the introduction among them of useful knowledge, and of religious and moral improvement, and in furtherance of the above objects, sufficient facilities ought to be afforded by law to persons desirous of going to and remaining in India. (Carson 1988: 411)

The passage of the clause had an immediate effect on East India Company's policy of awarding licenses to missionaries in India. The first CMS missionaries were given licenses to proceed to India at the end of 1813 (Carson 1988: 311). Most of the early missionaries focused their efforts on Hindus living in the Southern and Eastern parts of the country and had made "very little contact with the Muslims of the [Northwestern Indian] region until as late as 1830s" (Powell, 1993: 81). Powell argues that this had to do with perceptions of Muslims as unreasonable barbarians with whom a theological "impasse had long been reached" (Powell, 1993: 81) and also because of the desire to focus on the 'more primitive' majority Hindus who would be easier to convince of the obvious errors of their ways. This early optimism faded, however, as there were few Hindus willing to convert to Christianity and face social ostracism from their communities. The few missionaries who did make attempts to engage Muslim ulema and ashraf found that Muslims were convinced that they already knew everything they needed to know about Christianity. Early missionaries such as Henry Martyn and Joseph Wolff had trouble getting into the darbars of various Indian Muslim kings or getting Muslim scholars to engage in a *munazara*–a public religious debate–with them. The missionaries labeled this attitude 'hauteur' and 'arrogant' (Powell, 1993: 74). The CMS missionary Rev. T. P. Hughes argued that such an attitude was inherent in Islam:

> The special difficulties in the way of conversion of Mohammadans appear to arise from that self-conscious superiority and arrogance which is inherent power in Islam. Mohammadanism was never intended to be the religion of the conquered, and hence it is that the pious Moslems regards all other religionists, whether Jews, Christians, or fire-worshippers, as real objects of pity. The inspirations of the Koran and the divine mission of Mohammed are always taken for granted... the mind of the Moslem is so saturated with dogma that it instinctively repels any suggestions of inquiry, and consequently, in the majority of cases where an honest mind in Islam begins to inquire, he soon finds himself landed in the regions of practical atheism. (Hughes, 1879: 329)

The only way to break through the Muslim haughtiness would be to be provocative. Powell argues that this was the strategy adopted by Rev. Carl Gottlieb Pfander–the CMS missionary to Agra from 1841 to 1855. Pfander published three anti-Islam polemical works in Persian including "Mizan al-Haqq", "which remains until now the single most provocative Christian contribution to the Christian-Muslim polemical interchange" (Powell, 1993: 138). Pfander labeled *hadis* (traditions of prophet Muhammad) as 'gross fiction,' Islam as a 'system of falsehood,' Islamic theology a 'mire of error and superstition' and the prophet Muhammad as 'deluded' and 'false prophet' (Powell, 1993: 145-151). Immediately upon getting to Agra, Pfander sent Urdu translations of his books to Muslim ashraf and ulema of the city. Powell notes that “instead of ignoring them as before” the ulema responded to Pfander. This Powell argues showed that “some of the ulema were undergoing a change of opinion about the missionaries... and they were beginning to view their activities, and British rule generally as a growing danger to Islam” (Powell, 1993: 170). Powell argues that “the catalysis in this transformation from passive observance to active remonstrance, was the campaign launched by Carl Pfander in 1841, to engage ashraf Muslims in religious interchange” (Powell, 1993: 264). Sir Syed Ahmad Khan, a pro-British Muslim Indian civil servant, also blamed aggressive missionary preaching for the change in Muslim perceptions of missionaries. In his book, “On Causes of Rebellion of India,” he wrote:

> The missionaries, moreover, introduced a new system of preaching. They took to printing and circulating controversial tracts, in the shape of questions and answers. Men of a different faith were spoken of in those tracts in a most offensive and irritating way. In Hindustan these things have always been managed very differently. Every man in this country preaches and explains his views in his own mosque or his own house. If any one wishes to listen to him, he can go to the mosque or house and hear what he has to say. But the missionaries’ plan was exactly the opposite. They used to attend places of public resort—markets, for instance, and fairs, where men of different creeds were collected together—and used to begin preaching there. It was only from fear of the authorities that no one bade them be off about their business. In some districts the missionaries were actually attended by policemen from the station. And then the missionaries did not confine themselves to explaining the doctrines of their own books. In violent and unmeasured language they attacked the followers and the holy places of other creeds, annoying and insulting beyond expression the feelings of those who listened to them. (Khan, 1885: 42)

Maulana Rahmatullah Kairanawi–a Sunni *alim* and principle of a madrassa at Kairana, accepted Pfander’s challenge of a munazara debate. At the munazara, held 11-12 April 1854 at Agra’s Church Missionary Society School, Kairanawi was assisted by English speaking surgeon of Lucknow Dr. Wazir Khan while Pfander was assisted by his fellow CMS missionary Rev. Thomas Valpy French. After going through the traditional Islamic objections against Trinity, the discussion moved to Muslim allegations of *tahrif* (corruption) in the Bible. Like other evangelicals of the time, Pfander denied “any alterations in any manuscript or printed editions of any part of the Bible, apart that is from, any insignificant ‘copyist’ errors in any editions” (Powell, 1993: 243). Surprisingly, for the missionaries, the Muslim debaters cited “’evidence’ against the missionary claims from biblical commentaries, recent works of Biblical criticism, ecclesiastical and secular histories, and from a miscellaneous assortment of European reference works” (Powell, 1993:230-231). This seems to have forced, at least Rev. French and the presiding judge Mosley Smith, to admit to ‘various readings’ of the Bible. The debate ended in bitterness with both sides claiming victory and refusing to “sign the written accounts of the debate drawn up by the opponent, and both made allegations of deliberate falsification of the record” (Powell, 1993:255). Muslims perceived that they had ’taken fortress of Christianity’ and utilized, “every available vehicle of effective communication from *fatwa* to newspaper report... to broadcast the Muslim 'victory' over the missionaries" (Powell, 1993:264). Even though both Kairanwi and Khan were charged by the British authorities for their role in the 1857 mutiny, they both managed to escape to Mecca. Moulana Kairanawi was hailed as a modern day Saladin and invited by Ottoman Sultan to Istanbul to write a rebuttal to Christian missionaries. His two volume Arabic book "Izhar ul Haq" was published in 1864-65 and translated into Turkish, English, and French (Powell, 1993:295).

As discussed earlier, the missionaries became more cautious in their activities following the 1857 mutiny because of the perception by some British officials that fears of Christianization had caused some of the native soldiers to turn on their British masters. This does not mean, however, that there was any decrease in the number of missionaries. On the contrary, the number of missionaries continued to steadily increase throughout Northwestern India. According to Cox (2002), in the second half of nineteenth century, numerous Christian societies were active in Punjab (see Table 6).

> In the nineteenth century, Punjab encompassed much of what is now Pakistan and northwest India (Including Delhi). Although subject to Christian influences at various times, including well-documented Jesuit attempts to influence the Mughal court, this part of the world contained as a practical matter no Christians at all at the beginning of the nineteenth century. By the beginning of the twentieth century, however, the Punjab and its Himalayan hinterlands had been the scene of major efforts by missionary societies representing the principal Protestant denominations of both Great Britain and the United States: evangelical and high church Anglicans; the established church of Scotland; two competing groups of American Presbyterians; American Methodists; British Baptists; and the Salvation Army-as well as Roman Catholics (mostly Flemish speaking). Smaller missions followed in their train, and in the early twentieth century (the heyday of foreign mission work in India) more than thirty societies were at work alongside innumerable private individuals and small groups of locally based philanthropists and missionaries. (Cox, 2002: 2)

Table 6: A list of missionary societies that were active in Northwestern India in the late nineteenth century and early twentieth centuries compiled by (Cox, 2002: 2).

Presbyterian Church in the USA
Associate Reformed Presbyterian Mission
British and Foreign Bible Society
Baptist Missionary Society
Baptist Zanana Missionary Society
Central Asian Mission
Church of England in Canada Missionary Society
Church of England in Zanana Missionary Society
Cambridge Mission to Delhi
Christian Missions in Many Lands ("Open Brethren")
Chuch Missionary Society (Anglican)
Church of Scotland
Danish Pathan Mission

Continued **Table 6:** A list of missionary societies that were active in Northwestern India in the late nineteenth century and early twentieth centuries compiled by (Cox, 2002: 2).

Evangelical Society of the Pittsburgh Bible Institute
Female Education Society
Indian Female Normal School
India North West Mission
Methodist Episcopal Church North
Mission to Lepers in India
Moravians
National Missionary Society (Indian)
New Zealand Presbyterian Mission
Punjab Religious Book Society
Roman Catholic
Salvation Army
Seventh Day Adventists
Society for the Promotion of Christian Knowledge (Anglican)
Society for the Propagation of the Gospel (Anglican)
Unevangelized Fields Mission
United Presbyterian Church of North America
World Evangelization Crusade
Young Men's Christian Association
Young Women's Christian Association
Zenana Bible and Medical Mission

In the wake of the 1857 mutiny and the debacle of Pfander-Kairanawi debate, the Christian missionaries paused to take stock of the situation and to chart a new course more suited to the new environment in Northwestern India. The Punjab Missionary Conference was held in Lahore, at the end of 1862 and the start of 1863, to compare notes on the effectiveness of various strategies for preaching to Muslims, Hindus, and Sikhs. The first speaker in the opening session of the conference, Rev. John Newton of the American Presbyterian Mission in Lahore admitted that current missionary strategies may not have been effective:

> The preaching of Gospel in the Punjab, however, and indeed in most parts of India, in respect to its primary object, has hitherto met with very small success: for, though a wide impression may have been made, and doubtless has been made in favour of Christianity, few souls have actually been converted to Christ. What reasons can be assigned for this? And how can preaching to the heathen, in these parts, be rendered more efficacious? (Compilation-Committee, 1863: 4)

Answering his own question, he argued that the most important element of successful preaching involves, "giving no unnecessary offense to their prejudices; but seeking a conciliatory mode of address" (Compilation-Committee, 1863: 4). The use of 'controversy' in preaching was a major topic of discussion at the meeting .

Summarizing the discussion, Guenther notes that, "a number of missionary writers actually shared Ahmad Khan's criticism of bazaar preachers" (Guenther, 1998: 10). Rev. Newton continued to argue that Jesus's status should be the central point of preaching:

> Jesus, as the incarnate Son of God, endowed with all the attributes of a mighty Saviour, cannot be held too much to the view of the heathen...Let it be seen that he was indeed spotless, and benevolent, and self-denying, as well; as mighty, beyond comparison... The resurrection, too of Jesus should occupy a prominent place, as the crowning miracle of his incarnate state, as the great attesting seal of all his claims... (Compilation-Committee, 1863: 6)

Rev. J. H. Orbison, who followed Rev. Newton, agreed with him that the use of controversy "has been shown to be injurious in most cases" and should be banished "from Mission policy and practices" (Compilation-Committee, 1863: 26). He argued that a conciliatory start can be had by learning about the religious beliefs of the natives and by emphasizing the common ground between Christianity and native religions.

> 1. Every Missionary ought to be thoroughly acquainted with the languages and notions of the people. He ought to have a good knowledge of the books, and religious maxims, and prejudices, of the Hindoos and Mahomedans. For want of this, much time and breath are often spent in vain, by encouraging and prolonging disputations which might have been prevented, or at least finished, with a few words. Answers to many objections can be drawn from their own books and opinions.
>
> 2. The plan of dwelling on the subjects of agreement and points of contact, instead of those of repulsion and disagreement, would doubtless facilitate the good work, and tend to conciliate and win over hearers. (Compilation-Committee, 1863: 27)

Rev. Thomas Patrick, a CMS missionary to Peshawar from 1864 to 1884, admitted that the Muslim charge of corruption in the Bible – the central point of the Pfander-Kairanawi debate– is "the most plausible of modern objections" (Hughes, 1875: 22). Instead of focusing their energies on such controversial issues, he advised missionaries to focus on traditional Islamic beliefs about Jesus which he argued not only form a common ground between Muslims and Christians but can be exploited by missionaries prove the truth of Christianity.

> In dealing with Muhammadans the Christian missionary must not treat their system as though the views of Islam were precisely those of modern Socinians. Islam admits the miraculous conception of Christ, and that he is the ''Word'' which "God conveyed into Mary"; and whilst the other five great prophets are but "the chosen," ''the preacher," 'the friend," "the converser with," and "the messenger" of God, Jesus is admitted to be the "Spirit of God.'' He is the greatest miracle worker of all the prophets, and whilst Muhammad is dead and buried, and saw corruption, all Muslim divines admit that Jesus "saw no corruption," and still lives with a human body in Paradise. (Hugh, 1875: 265)

The new 'non-controversialist' missionary strategy that became dominant in the second half of nineteenth century in Punjab involved engaging Muslims in a friendly discussion by starting with the commonalities between Islam and Christianity and then proceeding to argue that Islamic sources themselves admit to Jesus' higher status and Muhammad's lower status. The genious of the idea was to convert the Muslim knowledge of Christianity from being a source of Muslim strength into a source of weakness that could be exploited by missionaries. They hoped that this new approach would finally allow them to break through the 'hardness' and 'bigotry' of the Muslim heart. Rev. T.J. Mayer of the CMS had a chance to witness the effectiveness of this strategy firsthand during a preaching session in Bannu in 1875.

> Gul Khan, the blind Hafiz is to be baptized, please God, on Sunday, i.e., if they do not murder him or poison him beforehand... I was sitting, on April 24th, in the verandah, after my English class, when Synd Hakim Shah, a member of the municipal committee here, came in to beg me not to teach his son the Gospel... I had got some way in my discourse, when in came the blind man; I referred the Synd to him for an answer as to what he had seen of the beauty and purity of the Gospels. Then followed a long and most interesting discussion. G.K. [i.e., Gul Khan] quoted passage after passage in Arabic from the Koran showing the authority of the book of Moses, the Psalms, and Gospel; the account of Christ in the Koran owning Himself as the Son of God, etc. The Synd sat astonished, and although he made several answers, it was evident his faith had gotten a shock from which it will not readily recover. G. K. then burst out into a splendid peroration on...the humanity only of Mohammed, and the humanity and divinity of Christ. It was a very pleasant time... They went away, and one of our Christians then told me that he wanted to be baptized. (Mayer, 1876: 284)

The new 'non-controversialist' approach to preaching appeared to have been more effective than the older 'controversialist' approach. In 1893, Rev. Dr. Imad-ud-din claimed that "there was a time when the conversion of a Mohammedan to Christianity was looked on as a wonder. Now they have come and are coming in their thousands" (Imad-ud-din, 1893a: 582). Robert Clark, head of CMS in Punjab, boasted that while there were virtually no Christians in Punjab prior to the establishment of the CMS mission in 1852, they numbered around thirty thousand by 1894.

So impressed was he by quick growth of vegetation in Punjab's fertile lands that he found allusions to it as the best way of illustrating the rapid expansion of the church in the late nineteenth century. The way, "a little twig soon grows into a great tree, if it receives water and care" he said the Christianity in Punjab has also grown quickly. "A little vine was then planted, which has taken root, and it is gradually spreading itself over the land" (R. Clark, 1883:4).[26]

> Two mission house were built in 1852... Our City School house was built in 1853... The Jandiala Mission was commenced, and a small house built in 1854... Two orphanage houses were built in 1855... The Lady Henry Lawrence Schools were established in 1856-58... The Native Church in Amritsar was built there in 1866... The City Mission House (where His Royal Highness the Prince of Wales received the Native Christians of the Punjab in 1876), the Native Pastor's House, the Christian Serai, and the Mission Room, Called Shamoun's Jhanda (the flag for Christ) were built

> in 1866 and 1867... The Batala Mission was established in 1866. The Native Church was twice enlarged in 1866 and 1875... and is being again enlarged in 1883... (Clark, 1883: 32).

He continued:

> The little sapling planted in 1852 has already become a great tree; and has thrown out many branches on every side. And the branches are growing, and are throwing out other twigs and shoots, which will themselves soon become branches; and their leaves are furnishing medicine and shade to many people; and their fruits are feeding many from the Tree of Life. (Clark, 1883: 46)

Rev. Clark emphasized that the establishment of missions in Punjab was not the result of a human plan but because of 'providence of God' who recognized the central importance of Punjab to spreading Christianity among to the entire Muslims world:

> If we accept the position in which God's providences have placed us, and try to realize the vast opportunities which He has given us, we have to consider what kind of agencies we require to fulfil these great responsibilities... with very few exceptions, there are absolutely no Christian Missions beyond us. We may travel eastward, northward, and westward, to the confines of China, to almost the Arctic regions, or to Palestine and Constantinople, without meeting (with the exception of the Moravian Missions in Lahoul, and a few scattered missionaries in Persia and Armenia), as far as we know, with any living Christianity at all. It is from out Punjab Frontier line, and with it as our basis of operations, that Christianity must advance onwards to the countries where it is yet unknown. (Clark, 1883: 26)

As Porter (2000) argues, Evangelicals saw Christian Europe's military success against Christianity's old nemesis Islam as a sign of the end times. They believed that fading of Muslims power in North Africa, the Middle East, and South Asia would allow them to finally put the Islamic heresy to an end and usher in the second coming of Jesus. Rev. French thus argued that "everywhere we find Mohammedanism... hotly and sorely pressed in a life and death struggle" (as quoted by Porter 2000: 116). At Church Missionary Society meeting in the 1880s, the vicar of Fareham opined: "The fifth horn in the vision of the ram and the he goat in the eighth chapter of Daniel is a symbol of the Mohammedan power, and that its time for practicing and prospering against the Prince of princes is now coming to an end... Certainly it is a sign of times that the crescent is waning before the cross." "Islam sees all her frontiers falling in" wrote William Arthur. Maj Gen Haig noted that, "days for Mohemmaden Antichrist are numbered. The disintegration of the Turkish Empire proceeds apace." In combination with the obvious decline in Muslim fortunes throughout the world and the rise of Christian Europe, they were convinced that Indian Muslims as well as Muslims around the world, would soon be converted to Christianity. Mass conversion of Indian Muslims, believed missionaries, would open the door to the conversion of Muslims throughout the Middle East (Porter, 2000). Rev. Imad-ud-din argued:

> There was a time of their ascendency when Mohammedans conquered many lands. That speedily went by. Since then nothing is to be seen in Islam, all the Mohammedan world over, but decay on decay... The reason of the backwardness of Mohammedans and their low estate in things religious, as well as things worldly, is the same–it is simply and solely teachings of Mohammed, and the foolish things that obtain among them. These render all progress hopeless... Mohammedans and others are now so utterly crushed and annihilated that they will not recover themselves until the day of judgment. (Imad-ud-din, 1893b: 583-584)

The Punjabi Muslim *ashraf* who in their earlier haughtiness had believed that since Islam was so superior to Christianity, no Muslims could ever abandon Islam, were now shocked by the rising wave of apostasy. As Rev. Siraj-ud-din pointed out, "the largest number of Christian converts from Islam in India are from the Panjab" " (Siraj-ud-din, 1913: 96). Far from seeing the 'non-controversial' missionary approach as less threatening than the previous missionary approaches, Muslims saw the dramatic expansion in missionary activity as part of a grand plan to convert Muslims around the world to Christianity. The use of militaristic and triumphant terminology by the missionaries also heightened the sense of danger. As high-identifiers, they genuinely feared an annihilation of their group unless something was done to confront the adversary. The missionary notions of the end of the world served to reinforce their own eschatological expectations about the imminent arrival of a savior who would defeat Islam's opponents, such as Christian missionaries. Savior of Islam was exactly what Mirza Ghulam Ahmad claimed to be. In the next chapter, we will learn more about Ahmad's family and the environment he grew up in and how that environment shaped his beliefs.

6.6 Chapter Summary

After the fall of the Mughal empire, Indian Muslims families with claims to ancestory from Islamic lands of Central Asia and Arabiaincreasingly thought of themselves as vanguards of Islam in the land of heathen Hindus. The initial loss of Muslim political and military power in the eighteenth century was seen by them as a temporary setback similar to many that had happened in the –a-millennia long history of Muslim power in India. However, as hopes of a quick recovery faded and a large number of Christian missionaries descended upon India (because of socio-cultural and political developments in Britain), Muslims increasingly took refuge in their eschatological beliefs about national saviors who were to appear at their greatest hour of need to save Islam from extinction and to restore the lost Muslim glory. Expectations of an imminent arrival of two end-of-the world figures were particularly rife among nineteenth century Indian Muslims. The first savior was to be Jesus, whom Muslims believe to have been a righteous prophet of God whose divine message had become corrupted into Christianity. Muslims believe that God raised Jesus to heaven alive to save him from the indignity of crucifixion. According to the shared Sunni Indian

Muslims beliefs, Jesus sits on the right hand of God awaiting his return before the end of time. The returned Jesus would work as a second-in-command to a Muslim reformer titled Mahdi (i.e., the guide). The two men would lead an army of Muslims to defeat all infidels and make the entire world Muslim. Christian missionary activity also played a role in making beliefs about Jesus prominent in the minds of Indian Muslims. Missionaries found that perceptions about Jesus provided them an opening to engage Indian Muslims who were seen to be the most resistant of all Indians to the Christian missionary efforts. Frustrated with the failure of their earlier proselyting efforts, missionaries had taken to telling Muslims that their own teachings showed that Jesus was superior to Muhammad. Thus, while Muhammad lay buried six feet deep in the ground, having died long ago, Jesus was still alive and sat on the right hand of God. This strategy resulted in thousands of Punjabi Muslim converting to Christianity raising fears of the extinction of Islam in the minds of Northwestern Indian Muslims. This common social threat was discussed with an increasing sense of alarm in elite North Indian Muslim literary circles.

7 Mirza Ghulam Ahmad

> In 1530, the last year of the Emperor Babar's reign, Hadi Baig, a Mughal of Samarkand, emigrated to the Punjab and settled in the Gurdaspur district. He was a man of some learning and was appointed Qazi or Magistrate over 70 villages in the neighbourhood of Qadian, which town he is said to have founded, naming it Islampur Qazi, from which Qadian has by a natural change arisen. For several generations the family held offices of respectability under the Imperial Government, and it was only when the Sikhs became powerful that it fell into poverty. Gul Muhammad and his son, Ata Muhammad, were engaged in perpetual quarrels with Ramgarhia and Kanahaya Misals, who held the country in the neighbourhood of Qadian; and at last, having lost all his estates, Ata Muhammad retired to Begowal, where, under the protection of Sardar Fateh Singh Ahluvalia (ancestor of the present ruling chief of the Kapurthala State) he lived quietly for twelve years. On his death Ranjit Singh, who had taken possession of all the lands of the Ramgarhia Misal, invited Ghulam Murtaza to return to Qadian and restored to him a large portion of his ancestral estate. He then, with his brothers, entered the army of the Maharaja, and performed efficient service on the Kashmir frontier and at other places.
>
> During the time of Nao Nahal Singh and the Darbar, Ghulam Murtaza was continually employed on active service. In 1841 he was sent with General Ventura to Mandi and Kalu, and in 1843 to Peshawar in command of an infantry regiment. He distinguished himself in Hazara at the time of the insurrection there; and when the rebellion broke out, he remained faithful to his Government and fought on its side. His brother Ghulam Mohaiudin also did good service at this time. (Griffin & Massy, 1890: 35)

Dard (1948) goes into more detail about the tragic events that befell the Mirza family of Qadian around 1802 when they were forced to leave Qadian and seek shelter in Begowal.

> ...the Sikhs of Ramgarh, Jassa Singh or his followers, found their way into Qadian through treachery, and the members of the family were all made prisoners. Everything was looted. Mosques and buildings were pulled down, and one of the mosques was turned into a Gurdawara, i.e., a Sikh temple, which can be seen to this day. The whole of the library containing a large number of valuable books was burnt to ashes. A number of people were killed, but the members of the family were spared and on a cold wintry night they were all expelled from Qadian. They had to leave the town by night, shivering with cold and overcome with grief. They repaired, shuddering and exhausted, to a village called Begowal, where... Sardar Fateh Singh Ahluvalia, an ancestor of the Maharaja of Kapurthala, treated them with kindness, and extended to them a practical sympathy... In consideration of their sad plight he granted them an allowance... The family remained there for about 16 years... Ata Muhammad was poisoned in 1814 by his enemies. His son Mirza Ghulam Murtaza was quite young at that time, but he brought his father's body to Qadian, so that he might be buried in the family cemetery and thus the ancestral connection with Qadian and the claim over the estate might remain intact. The Sikhs opposed this; but the local population, consisting of humble people, were very much excited and the Sikhs, fearing open rebellion, had to give in. (Dard, 1948: 12-13)

Mirza Ghulam Murtaza was finally able to bring his family back to Qadian after the defeat of Ramgarh Sikhs at the hands of Ranjit Singh. In the late 1830s, Ranjit Singh restored Murtaza's sovereignty over five villages around Qadian. Despite this gesture, the enmity between the Sikhs and Mirzas was far from over as, according to Dard, "the Sikhs made, in their last days, an abortive effort to kill Mirza Ghulam Murtaza and his brother, Mirza Ghulam Muhyuddin, who were confined by them in Basrawan, near Qadian; but they were soon rescued by their younger brother, Mirza Ghulam Haidar" (Dard, 1948: 15).

In the late 1830s, just as the Sikh rule was finally being brought to an end by the British, Mirza Ghulam Murtaza and his wife Charagh Bibi had their second son[27]. They named him Ghulam Ahmad (i.e., the slave of Ahmad–Ahmad being another name for prophet Muhammad). Even though Mirza Ghulam Murtaza had served in Ranjit Singh's army, he supported the British in their fight against Ranjit Singh's descendants. The British reciprocated by granting Murtaza a pension of Rs. 700. When the British needed help to put down the 1857 mutiny, Murtaza raised a small army of 50 soldiers including his elder son, Mirza Ghulam Qadir, to support the British. They served as part of General Nicholson's militia and helped the British retake Delhi from the mutineers.

As were other noble men of his time, Murtaza was not just a military leader but also a *hakim* (i.e., a doctor of traditional medicine) and a poet. He also had some interest in religion as he spent considerable resources to purchase land and construct a large mosque in Qadian. Since none of Murtaza's writings have survived, we know little about his precise religious views. We can, however, attempt to infer his thoughts by analyzing his actions that have been narrated by Ahmad and by Ahmad's biographers. For instance, his selection of teachers to homeschool young Ahmad is a case in point. Murtaza hired three teachers including a Hanafi Sunni named Fazal Ilahi, an Ahl-e-Hadith alim named Fazal Ahmad of Sialkot, and a Shia alim named Gul Ali Shah of Batala. These would hardly have been the choices of a traditional Hanafi Sunni who believed in strict adherence (*taqlid*) to a particular mazhab. Instead, they suggest a non-denominational approach such as that advocated by Ahl-e-Hadith leaders of nearby Delhi. Here's how Ahmad describes his early education.

> When I was six or seven a Persian-speaking was pressed into my service as a teacher. He taught me Holy Quran and a few Persian books. This respectable man's name was Fazl Ilahi. And when my age was about ten-years then an Arabic speaking Moulavi Sahib, whose name was Fazl Ahmad, was appointed for my instructions... The praiseworthy Moulavi Sahib was a religious and respectable man. He taught me with much attention and hard work. And I studied some books of *sarf* and *nahv* from him. And when I was seventeen or eighteen, I happened to have the chance to study under another Moulavi Sahib for a few years. His name was Gul Ali Shah. He was also pressed into service by my father to teach me in Qadian. From the latter Moulavi Sahib, I acquired as much knowledge of *nahv*, logic, and traditional medicine etc. as Allah had intended. And I read some books of traditional medicine from my father who was a great physician. (Ahmad, 1898: 149-150)

There is also no mention by Ahmad or any of his biographers of any visits by his family to any Sufis or their *mazars* (shrines) despite the fact that this was a common practice among traditional Hanafi Sunnis in rural Punjab. Nor is any affiliation with a traditional Sufi order ever mentioned. There were several tariqa Sufis in towns all around Qadian and for an ashraf family with attachment to a Sufi tariqa, it would have been customary to send their son to be trained by a Sufi and be initiated into the order.

All of Ahmad's Ahmadi biographers agree that Ahmad moved to the family's second home in Batala to continue his study (probably around 1857). Shahid says, "even though Moulavi Fazal Ilahi and Moulavi Fazal Ahmad lived in Qadian till the end of their employment, but Moulavi Gul Ali Shah moved to Batala after a short stay in Qadian" (Shahid, 1958: 55). Thus Ahmad had to move to the nearby town of Batala to continue his studies. During this time, he became friends with a fellow ashraf disciple of Gul Ali Shah named Muhammad Hussain Batalavi. This was before Batalavi moved to Delhi to study under the famous Ahl-e-Hadith alim Syed Nazir Hussain. Ahmad and Batalavi's friendship was both social and intellectual and lasted for more than three decades. Years later, Batalavi publicly wrote about this friendship in his magazine Ishat-us-Sunnah:

> Few of my contemporaries know the author of Brahin-e-Ahmadiyya as well as I do. Not only are we from the same part of the country but in our young age (when we studied *Qutbi* and *Sharh Mulla*) we were also class mates. Since then he and I have continuously written to each other. Therefore, to say that I am very familiar with his circumstances and thoughts won't be a lie. (Batalavi, 1884: 176)

Ahmad also valued Batalavi's friendship just as much. In an 8 September 1887 letter to Batalavi, he wrote:

> I know for sure that you have had true fondness (for me) and my heart attests that you are engaged in supporting my activities with your pen and tongue only due to the enthusiasm of true endearment. (Ahmad, 2008: 305)

Then again in a 5 October 1887 letter, Ahmad wrote:

> I believe that compared to most ulema of the day, and if you don't mind, in some divine endeavors even in comparison with Moulavi Nazir Hussain, you are better… I am fond of you because of your inner piety. (Ahmad, 2008: 312)

The relationship between Ahmad and Batalavi also extended to their families. Batalavi's father was a *mukhtar* (legal assistant) and Ahmad's father seemed to be in constant need of legal help on account of his large estate. There was always a serf who refused to submit the due share of the crop or a neighbouring farmer who encroached on Murtaza's water or land rights. Some of the legal proceedings took place at the Batala Magistrate Court and hence Ahmad's family needed to acquire a second home in Batala. Ahmad's letters make clear that Ahmad not only knew Batalavi's father but had also interceded on Batalavi's behalf to prevent Batalavi's father from taking adverse actions against his son presumably due to a shortcoming on Batalavi's part. Ahmad's letters also make reference to multiple meetings between their fathers.

As was the family tradition, Ahmad was married at a young age (in 1854 when he was probably 15) to the daughter of his maternal uncle, Mirza Jamiat Beg. He had two boys, named Mirza Sultan Ahmad and Mirza Fazal Ahmad, with his first wife. There is little mention by Ahmad or by any of his biographers of Ahmad's first wife or their first two sons. What was the state of their relationship? What was his wife's name? Did she and her sons move with him to Batala? What religious beliefs did she and her sons hold? Sir Griffin and Col. Massy in their 1890 description of the Chiefs of Punjab mention Mirza Sultan Ahmad as having been adopted by Ahmad's childless older brother Mirza Ghulam Qadir. None of his biographers (or Ahmad himself) describe how Ahmad felt about having to give his son up for adoption. We know that Ahmad's first family did not join Ahmadiyya Muslim Jama'at during Ahmad's lifetime and seemed to have remained non-denominational reformed Sunni Muslims similar to their grandfather and uncle.

Ahmad grew up as the youngest child of the family (other children were born after him but they died young). His brother Ghulam Qadir may have been as much as a decade older. As they grew up, Qadir shared their father's interests and inclinations more closely than Ahmad did. This is what may have led Ahmad's father to prefer Qadir over Ahmad. He is reported to have described Qadir as a 'capable' helper and Ahmad pejoratively as a 'disabled' individual "who has given up all relations" and as someone who is "dead in this life" (Ahmad, 1939: 27-28). Murtaza was deeply disappointed in his younger son because he seemed to have withdrawn from the world and closeted himself in the family mosque. Ahmad talks about often being the "aim of my father's anger" (Ahmad, 1898: 156). Ahmad was much closer to his mother. While Qadir accompanied his father in various military missions in various parts of North India, Ahmad stayed home with his mother.

> Ghulam Kadar was serving in the force of General Nicholson when that officer destroyed the mutineers of the 46th Native Infantry, who had fled from Sialkot, at Trimughat. General Nicholson gave Ghulam Kadar a certificate, stating that in 1857 the Qadian family showed greater loyalty than any other in the district... Ghulam Kadar was always active in assisting the local authorities, and held many satisfactory certificates from officers connected with the administration. He enjoyed a reduced pension of Rs. 180 per annum. (Griffin & Massy, 1890: 50)

General Nicholson himself acknowledged Ghulam Qadir's assistance in a letter to him in August 1857.

> You have... helped the Government at your own expense with 50 sowars and horses, therefore, in recognition of your loyalty and bravery this parwana is addressed to you, which please keep with yourself. The Government and its officials will always have due regard for your services

and rights, and for the devotion you have shown to the Government. After the suppression of the insurgents I will look to the welfare of your family. I have also written to Mr. Nisbet, Deputy Commissioner, Gurdaspur, drawing his attention to your services.' (Ahmad, 1898: 6)

Thus it was Qadir, not Ahmad, who came to inherit his father's mantle as the Chief of Qadian. Griffin and Massy describe Mirza Sultan Ahmad as the successor to Qadir bypassing Ghulam Ahmad. It may have been in response to Gen Nicholson's recommendation that Qadir was hired as a police sub-inspector by Mr. Nisbet. Ahmad's father also asked his contacts in the government for a job for Ahmad. But unlike his older brother who was offered an officer's position to serve in the close by city of Gurdaspur, Ahmad was only offered the lowly position of a court clerk in the relatively far away city of Sialkot. Similarly, while no expense had been spared for Qadir's lavish wedding ceremony, Ahmad's wedding was kept a low-key low-budget affair.

During his time in Sialkot, Ahmad seemed to have made a concerted effort to change his circumstances and to acquire the skills needed to succeed in worldly affairs. He convinced one of his fellow clerks to teach him English and he studied for months to write the *mukhtar* (legal assistant) exam. He failed in both of his efforts. His admirers argue that Ahmad failed in his worldly education because he was too busy studying Quran instead of studying his textbooks (Ahmad, 1939: 44), while his critics argue that he lacked the required intellect (Amritsari, 2002).

During the time he spent away from his family in Sialkot, Ahmad also felt free to pursue his own interests in religion. He met with Sufis, Christian missionaries, in addition to Muslim reformers. Ahmad may also have been able to reconnect with his Ahl-e-Hadith teacher Fazal Ahmad and his son Mubarak Ali who was the Imam of the Jamia Masjid (i.e., central mosque) of Sialkot. Ahmad also frequented the esteemed Naqshbandi Sufi Pir Shaikh Mahbub Alam to learn about Sufism. While it is not clear whether or not Ahmad ever took a formal *bayat* (oath of allegiance) at the hands of Alam, scholar of religion Adil Khan argues that "Alam may still have served as a spiritual guide for Ghulam Ahmad all the same, irrespective of whether Ghulam Ahmad was initiated into the Naqshbandi order" (Khan, 2015: 29). Because of their keenness to emphasize divine origin for Ahmad's ideas, Ahmad's Ahmadi biographers are loath to admit to any training he may have received from the ulema. Khan argues that "Ahmad's biographers appear to have consistently concealed the names and religious affiliations of Muslims capable of influencing his mission in any way" (Khan, 2015: 30). Ahmad's close ideological affinity with Ahl-e-Hadith ulema is seen particularly troublesome by Ahmadis because of the opposition that Ahmad subsequently faced from his former Ahl-e-Hadith friends. Thus, Ahmad's three-decade-long friendly relations with Muhammad Hussain Batalavi are glossed over while enmity during the last fourteen years of Ahmad's life are dwelled upon at length. Batalavi is often described by Ahmadis as Ahmad's nemesis and Ahmad's *Abu-Jahl* (a reference to the 8th century Meccan opponent of the prophet Muhammad).

Perhaps, because of the reputation that Syed Mir Hasan came to enjoy because of having been a teacher to Allama Muhammad Iqbal, Ahmadis discuss at length the intellectual friendship between Hasan and Ahmad during Ahmad's stay in Sialkot. Hasan was well-known for his reformist ideas and his support of Sir Syed's Aligarh movement. Hasan describes that other Muslims who had an influence on Ahmad during his Sialkot days were also reformers.

> Since Mirza Sahib liked to debate Christian clerics, Murad Beg who used the *nom de guerre* of Mirza Shikasta (and later on Muwahid) and was a resident of Jallandhar, told him, "Syed Ahmad Khan has written an exegesis of Torah and Injil. If you write to him it may be helpful." Thus Mirza Sahib wrote a letter to Sir Syed in Arabic. (Shahid, 1958: 97)

Discussions with these reformers and the relative isolation from his father's influence may have allowed Ahmad to develop his own independent belief system.

Ahmad only stayed in Sialkot for four years and hastily left the city with his father's messenger who came to inform him about his mother's illness in 1868. When Ahmad reached Qadian, he found out that his mother had already passed away. So worried had been Ahmad's father about Ahmad's reaction that he had instructed the messenger not to tell Ahmad about her death. Ahmad was so stricken with grief that his father didn't have the heart to send him back to Sialkot. Ahmad wrote a letter to his father describing his thoughts at this time.

> My revered father—May you live long and in peace. I wish to say that all around me I see, in towns and country, an epidemic that appears to separate friends from friends, and relations from relations. Not a year passes but that a calamity inflicts suffering on the world. In view of this, my heart has grown cold towards all around me, and I have turned pale out of fear. I often read the following verse of Shaikh Muslih-ud-Din Sa'di Shirazi, and shed tears of regret and grief:
>
> مکن تکیه بر عمر ناپایدار
>
> مباش ایمن از بازی روزگار
>
> Also the following verse from the work of Farrukh Qadiani[28] saddens my heart:
>
> بدنیاۓ دوں دل مبند اۓ جواں
> کہ وقت اجل میر سد نا گہاں
>
> Hence I wish I could spend the rest of my days in some solitary corner, drawing myself away from the company of men and busying myself with the remembrance of God. Perchance I may succeed in making up for lost time, and may be able to make amends for all my past:
>
> عمر بگذشت و نماند است جز از گا مے چند
> بہ کہ در یاد کسے صبح کنم شامے چندا
>
> There is no stability in this world. All life is transient.

On his return from Sialkot, Ahmad describes the circumstances in which he came to reconnect with his friend Muhammad Hussain Batalavi who was now a full-fledged sanad-holding Ahl-e-Hadith alim.

> In 1868 or 1869, when Moulavi Abu Saeed Muhammad Hussain Batalavi, who at one time was also my classmate, came back to Batala after having become a moulavi. The people of Batala were bothered by his ideas. A person pressured this humble servant into having a debate on some controversial issue with the praiseworthy Moulavi Sahib. Therefore, on his insistence, this humble servant accompanied that person that evening to the praiseworthy Moulavi Sahib's house. We found him along with his father in a mosque. In summary, this humble servant after hearing Moulavi Sahib's learned that there was no objectionable excess in Moulavi Sahib's speech. Therefore, I gave up the debate for God's sake. (Ahmad, 1884: 520:521)

That chance meeting between Batalavi and a twenty-something year old Ahmad turned out to be momentous occasion in Ahmad's life. It was on that fateful day that Ahmad received some of the most awesome revelations that he ever received.

> That night the powerful God through his revelation pointed to the same abandonment of debate and said that, "your Lord is pleased with this action of yours and he will award you many blessings. So much so that kings will seek blessings from your clothes." Then I was shown those kings who were riding horses. Since I had adopted humility and abasement purely for God and His prophet, therefore, that Absolute Benevolent did not want to leave it unrewarded. Thus think and reflect.
>
> Then He said that the benefit of God's blessings is that they can be used to cure people's afflictions. Those who are pure of the heart will be rightly guided through your sayings... After these revelations, some Farsi, Urdu, and an English revelation happened... "The days (sic) shall come when God shall help you glory be to this Lord God of earth and heaven." (Ahmad, 1884: 521)

The ideological compatibility between Ahmad and Batalavi suggests that despite the fact that Ahmad never received a sanad from an Ahl-e-Hadith alim, the knowledge that he had acquired, since the days of their co-discipleship of Gul Ali Shah, had led him to ideas not very dis-similar from those being advocated by the Ahl-e-Hadith ulema. The divine sanction of Ahmad's agreement with Batalavi's Ahl-e-Hadith doctrine brought Ahmad and Batalavi even closer together. Batalavi went on to have a major influence on Ahmad's life for the next couple of decades.

Having had to sell their own Batala house to pay their legal fees, Ahmad took to staying at Batalavi's house whenever he visited Batala (Ahmad, 1939: 27-28). Batalavi may also have introduced Ahmad to a fellow student of Syed Nazir Hussain, named Moulavi Abdullah Ghaznavi. Ghaznavi had been expelled from his native Afghanistan because of his Ahl-e-Hadith views and he had had to move to Punjab to save his and his family's life. He settled in the village of Khayrdi near Amritsar. Ahmad and Ghaznavi seemed to have hit it off from their first meeting. Ahmad frequently visited Khayrdi to meet Ghaznavi and the two became very close. Khan argues that Ahmad may have learned a lot more from Ghaznavi than Ahl-e-Hadith doctrine.

> On a visit to Ghaznavi's village of Khjayrdi, near Amritsar, Ghulam Ahmad requested the maulana for special prayers concerning an undisclosed matter. Upon receiving this request, Maulana Ghaznavi immediately went home and began to pray for Ghulam Ahmad. In the coming days after returning home to Qadian, Ghulam Ahmad received a letter from Ghaznavi relating a slight variation of the following Qur'anic verse as a revelation which he had seen in a dream: "you are our Protector, so help us against the disbelievers (anta mawlana fa'nsurna 'ala 'l-qawm al-kafirin)" Maulana Ghaznavi interpreted the revelation to mean that Allah would help Ghulam Ahmad with his predicament, similar to the way in which Allah helped he companions of the Prophet Muhammad through various tribulations. The revelation, however, was almost identical to the last verse of Sura al-Baqara (2:286). An overwhelming number of Ghulam Ahmad's revelations have repeated Qur'anic verses, similar to this revelation of Abdullah Ghaznavi. In this light, it would be interesting to see how frequently other recipients of divine revelation have repeated portions of the Quran and claimed it as their own. If this format is unique, then perhaps it was first observed by Ghulam Ahmad in the revelations of Abdullah Ghaznavi.
>
> On a separate occasion, Ghaznavi saw a vision in which he described a light (nur) descending upon Qadian, but his children were being deprived of it. (Khan, 2015: 32)

The religious belief system that Ahmad was developing through independent readings and discussions with Ahl-e-hadith ulema, seemed indistinguishable from mainstream Ahl-e-Hadith doctrine. Like Ahl-e-Hadith, Ahmad believed that Quran and authentic Ahadith were the only authoritative sources of Islamic doctrine and that issues unaddressed by them should be resolved through ijtehad rather than blind taqlid of traditional jurists. Ahmad was also vehemently opposed to taqlidi cult of saints and mazars. Like mainstream Ahl-e-Hadith, he also rejected the "excessive rationalization" of Sir Syed Ahmad Khan. Having become convinced of the rightness of the Ahl-e-Hadith doctrine, Ahmad started preaching his ideas to those around him including his father.

To read Ahmad describe his father, it seems as if the old man had two completely different personalities. On the one hand, we see a wildly successful man of the world: in this version, Mirza Ghulam Murtaza is a brave military leader who has nothing but success on the battlefield, he is a successful hakim who can cure kings and ordinary men alike with serious illnesses that don't seem to have a cure, and he is a feudal lord who successfully earns hundreds of rupees from his landholdings and holds court with other feudal lords and British officers. On the other hand, Ahmad talks about Mirza Ghulam Murtaza as a sad poet and a wannabe Sufi who regretted having spent all his life in futile attempts to restore family's lost estates:

> My father was often sad and lost because of his failures. He had spent some seventy thousand rupees in legal expenses but their result was failure. Because our ancestor's villages had long passed out of our possession so their return was wishful thinking. Because of this failure, my father led a life of pain and was deeply buried under a torrent of sadness and restlessness. Watching these circumstances gave me a chance to make a virtuous change. Because my father's bitter life gave me lessons of a generous life that is free of worldly immorality... He often said that if he had exerted as much effort for religion as he had exerted for this foul world, he would have been a Qutb or Ghous of his time... Once my father described this dream that he saw the Holy

> Prophet[sa] coming in state to his house. He ran to receive the Holy Prophet[sa] and thought of offering *Nazar*. When, however, he put his hand into his pocket, he found that he had only one rupee and that it was a counterfeit coin. This brought tears to his eyes. At this point he awoke. This he interpreted to mean that the love of God and His Prophet[sa], mixed with the love of this world, is nothing better than a false coin. He used to say that similar to me, the last part of my father's life was also spent wallowing in sadness, pain and bitterness. Everything he touched turned into failure. He used to quote this verse of his father i.e., my grandfather...
>
> Whenever I make a plan, fate laughs at me. This sadness and pain had much increased in his old age. Because of this thought, my respected father built a mosque in the center of this town and willed that he be buried in a corner of the mosque so that his ears could hear the name of God and who knows that may save me on the day of judgement. (Ahmad, 1898: 156-159)

Here we get another clue to the religious belief system of Ahmad's father. It seems that his father not only believed in the hierarchy of Sufi offices but also believed that one could attain progressively higher statuses through spiritual excercises. Ahmad rejected the worldly part of his father's legacy that was embodied by his older brother, while claiming to fulfil his father's unmet spiritual aspirations of wanting to become a "Qutb or Ghous of his time." Ahmad claimed that spiritual achievements were an essential part of his family's long legacy.

> During the early age of the Sikhs, my great grandfather Mirza Gul Muhammad was a famous and well-known Lord who had eighty five villages... He was so big hearted and generous that even in this difficult time, he awarded many villages to various displaced Muslim lords... about a hundred ulema, sulaha, and Hafiz of Quran inhabited his court and were awarded scholarships. God and his prophet's name was frequently mentioned in his court. None of his servants and relations ever missed prayers. Even the women who ground flour also performed the five ritual prayers as well as the optional predawn tahajjud prayer... the Afghans surrounding Qadian used to call it Mecca because during that tumultuous period, this was a blessed refuge for every Muslim. And whereas infidelity, anarchy, and cruelty could be seen everywhere else, the fragrance of Islam, virtue, and justice came from Qadian. The late Mirza Gul Muhammad was one of the respected shaikhs of his time, and capable of performing *karamah* (i.e., miracles) and *khawariq* (i.e., apparitions). Many men of God, reformers, and distinguished persons had gathered in Qadian. And even more strange is that many of his *karamah* (miracles) are so well-known that a large group of Islam's opponents have also attested to them... thus I've heard from my father that a Mughal minister came to visit him and said... had I known... I'd have made an effort to put you on the royal seat in Delhi... During the high time of his sickness, the doctors agreed that if you use alcohol for this disease, it'd help but they lacked the courage to suggest it. Eventually, some of them meekly suggested it. He [Mirza Gul Muhamad] responded by saying that if God wanted to cure me, he would have produced many other medicines, I don't want to use this polluted thing. I have accepted God's decision. A few days later, he died from this illness. Death was his destiny but his virtuous path became memorable for the fact that he preferred death over alcohol. (Ahmad, 1898: 137-142)

When his father died in 1876, Ahmad considered his older brother to be his caretaker. Because Qadir lived in Gurdaspur, he was not always able to take care of Ahmad as their father had done. However, similar to their father, Qadir was also unhappy with

Ahmad for his refusal to help out in the family business. Once, when Ahmad asked him for money to buy some newspapers, Qadir wrote back, "this is wasteful. You don't do any work and just read books and newspapers" (Ahmad, 1939: 66).

The books that Ahmad spent all his time reading included exegesis of Quran, Hadith and works by Sufis such as Ibn-e-Arabi and Abdul Qadir Jilani as well as books by reformist Muslims such as Sirhindi and Waliullah who condemned various Sufi excesses. The periodicals he avidly read included Urdu newspapers such as *Safir-e-Hind* of Amritsar, and *Akhbar-e-Aam* of Lahore, the Ahl-e-Hadith magazines such as *Manshoor-e-Muhammadi* and *Ishat-us-Sunnah* edited by Batalavi, and Christian mouthpieces such as *Nur Afshan* of Ludhiana (Shahid, 1958: 119). These were communal newspapers that historians believed played a critical role in raising communal awareness and tensions in the nineteenth century India.

While Ahmad may have met Christian missionaries before his move to Sialkot neither he nor any of Ahmad's biographers mention it. This may have been because the Christian missionary activity hadn't ramped up in Batala/Qadian area before the 1860s. There was no established mission in Batala until CMS established one there in 1866. The Church of Scotland established a mission in Sialkot a decade earlier in 1856 (Melton & Baumann, 2010: 696). During Ahmad's stay in Sialkot from 1864 to 1868, his biographers mention several meetings and debates that Ahmad had with Christian missionaries based there.

> At Sialkot Ahmad came in touch with Christian missionaries. He held several debates with them. Elisha (Swift?), a native Christian missionary, who lived to the south of Hajipura, once had a discussion with Ahmad... The Rev. Butler, M. A. was a very learned Christian missionary and he had arguments with Ahmad on several occasions. (Dard, 1948: 48-49)

Ahmad was given a copy of the Urdu translation of the Bible during his stay in Sialkot. He read it carefully highlighting various lines and writing footnotes. He wrote another letter to Sir Syed Ahmad Khan who had announced that he was writing an exegesis of the Bible from a Muslim perspective (a project that he never completed). According to Dard, his friends in Sialkot included Nasrullah, a Christian convert who taught at the Scotch Mission School (Dard, 1948: 50). One of Ahmad's Muslim friends, Murad Beg, who later called himself 'muwahid' (perhaps a clue to his Ahl-e-Hadith faith who were known as Muwahideen) and was "fond of debating missionaries. He would frequently visit Mirza Sahib to seek answers to various questions" (B. A. Ahmad, 1939: 97). One of Ahmad's future followers, an Ahle-e-Hadith named Basharat Ahmad, recalling the tactics used by Christian missionaries in Sialkot in the late nineteenth century, wrote:

> Those days, I used to live in Sadr Bazar Sialkot. Since I studied at the Scotch Mission High School in Sialkot City and was sick of the daily objections by missionaries. I used to be embarrassed by the missionaries' translation of the Quranic verse يعيسى انى متوفيک ورافعک الى as [Jesus's] having been physically fully lifted to the heavens. I used to say, "why didn't Allah lift our Holy Prophet

> peace be upon him to the heavens?" He embarrassed us in front of the missionaries by lifting Jesus alive to the heavens. (Ahmad, 1939: 331)

Thus it may have been in Sialkot that Ahmad first came to hear of the Christian missionary's new non-controversialist approach of using traditional Islamic sources to ease new recruits towards basic tenets of Christianity. Khan also emphasizes the importance of Ahmad's interactions with Sialkot Christians in developing Ahmad's worldview.

> The Christian missionaries of Sialkot provided new prospects for religious dialogue with which Ghulam Ahmad was unfamiliar in Qadian. This exposure opened up new modes of thought for Ghulam Ahmad in his youth and enabled him to debate eschatology and salvation in an endeavor to prove the superiority of Islam as a religion. The exchanges also provided Ghulam Ahmad an opportunity to improve his communication skills by articulating his views, finessing his arguments, and formally expressing his beliefs–both verbally and in writing–for the first time. These discussions were beneficial in many ways, especially since Ghulam Ahmad was still an amateur theologian, whereas his opponents were more experienced and better educated missionaries. His encounters with Christian missionaries facilitated a second period of spiritual growth, which enabled his thought to mature... (Khan, 2015: 28)

By the time, Ahmad moved back to Qadian, the CMS mission in nearby Batala had been well established and missionaries from there and Amritsar (e.g., Rev. Rowland Bateman) were visiting Qadian to proselytize (Ahmad, 1939: 76). On 6 March 1873, the American Mission in Ludhiana launched a weekly newspaper in Urdu to, "discuss all matters in an unprejudiced and unbiased manner." Nur Afshan was offered for the low price of 1 Ana per month ("lowest priced newspaper in the country" claimed the third issue) so that "poor people who are fond of reading papers but unable to do so because of the high cost of newspapers" can get the latest news (Nur Afshan, 1873). The newspaper made it possible for Urdu speakers such as Ahmad who could not read or write English to not only learn about latest advances in European science and technology (the 1894 issue reports the invention of "self-driving vehicles") but also directly hear the Christian message from the missionaries.

Nur Afshan epitomized the new non-controversial approach to Islam increasingly being adopted by Christian missionaries in Punjab in the late nineteenth century. Indeed the first eighteen issues of Nur Afshan did not contain a single article with a traditional pro-Christian or anti-Muslims message. Instead the magazine criticized moral evils of lying, prostitution, "using false weights," "lack of education," and "multiple marriages" from a common sense perspective. But in a sign of the hyper-partisan times, one of Ahmad's friends from Sialkot days, the above mentioned Mirza Muwahad perceived the attack on "multiple marriages" as an implicit criticism of Islam and responded with an article in *Manshoor-e-Muhammadi*. The 10 July 1873 issue of Nur Afshan responded to Muwahad as follows.

> Mirza Muwahad has written an article in issue number 10 of Manshoor-e-Muhammadi dated 8 Rabiul Awal. In the above mentioned article, Mirza Sahib has used pejorative words about the paper and comic words about Christian clerics. Well, this is not so shocking because when one is overtaken by anger one's wisdom leaves him. (Nur Afshan, 1873: 74)

Perhaps assuming that if an article that does not even mention Islam or Christianity is going to be perceived as an attack on Islam then one may just as well write articles explicitly comparing the two religions, the 10 July 1873 issue also featured an article titled "on the topic of intercession (*shifaat kay bayan mein*)." This was the first Nur Afshan issue to do so.

> How should one relate to God? This is a serious question that demands attention. Quran has one answer to this question. See Surah Baqra Rukoo 1-2, *ahsinu innallah yuhibul muhsineen* meaning do deeds because Allah befriends the virtuous. Next, Surah Maida Rukoo 10 presents some conditions for salvation. It says man amana billahil yomil akhir wa amila salihan fala khauffun alaihim meaning he who believes in Allah and performs good deeds therefore will not have any fear...Muslims consider Muhammad to be an intercessor and Christians also believe in the need for intercession and so do Hindus. The question is what are the characteristics of an intercessor?... My friends, I have not heard of anyone else besides our Lord Maseeh who has sacrificed for ever deliverance... This claim also does not fit Muhammad Sahib who only came to guide people. But our Lord Maseeh made this clear claim that I have come to save sinners. Therefore, how should we relate to God? The answer is through the intercession of Maseeh Issa because he has paid our debt. And on this basis, he has the capability to intercede on our behalf. Because he is our friend, our bailor, and our intercessor. And Allah ta-allah will never refuse His intercession of humanity. (Nur Afshan, 1875: 178)

After publishing eighteen issues without any mention of Christian-Islamic debates, Nur Afshan began to regularly feature articles such as the one quoted above. These articles were filled with quotations from Quran and Hadith. Nur Afshan used the traditional Islamic sources of authority to support its argument that Quran did not nullify the older holy books of Torah and Injil. Instead, they argued that Muhammad was a holy man whose mission was to convey Gospel's message to the Arabs. Through Quran, God (using the Islamic term "Allah" to refer to God) they said, has emphasized a high status for Jesus (using the traditional Islamic term "Hazrat Issa" to refer to Jesus) as a miracle-worker born of a virgin who was lifted to the heaven alive and sits on Allah's right hand waiting his return at the end of times. By extensively quoting ahadith, they argued that Muhammad (referring to him in respectable terms "Muhammad Sahib") had only claimed the status of a humble man for himself. They pointed out that Muhammad had refused to show miracles when demanded by his opponents and he died a natural death like all other men.

The ramped-up Christian activity in Punjab was starting to bear fruit. Reports of conversions to Christianity became a regular feature of Nur Afshan. For instance, 3 June 1875 issue contained this report titled "Growth of Christian Religion (taraqqi-e-Din-e-Issavi)":

> Because of God's blessings, the true religion is growing daily. The preachers of the true religion are enlightening inhabitants of Islands of the north and south. Good news for Christianity comes frequently. For instance, we find out from *Kokab-e-Issavi* of Lucknow that in the south of India, Church Missionary Society has been working for a long time. Thousands of people have entered the church from those areas... The total number of people in that mission is forty six thousand six hundred and twenty. In addition, another thirteen thousand people are seeking to join. (Nur Afshan, 1875: 178)

In a complete reversal of the situation a half century earlier when Muslims, believing that they had an upper hand, had avoided Christian missionaries, were now seeking debate with them. Ahmad's biographers say that missionaries made an effort to avoid Ahmad, who was eager to try his newly-rehearsed arguments during their Qadian visits. Ahmad also shared the lessons that he was learning through his interactions with Christians with his fellow Muslim debaters. His Ahmadi biographer, B. A. Ahmad mentions one such debater, named Mian Nabi Baksh who would visit him in Qadian to tell him about the latest Christian objections against Islam. "Hazrat Sahib would write down all the objections of non-Muslim critics... he had collected three thousand objections" (Ahmad, 1939: 75-76). After listening to the objections gathered by visiting Muslim debaters, Ahmad would offer various responses and ask the Muslim debaters to rehearse their arguments over and over again, "so that you don't run into any difficulties" at the time of the actual debate with missionaries (Ahmad, 1939: 75). Ahmad would play the role of an aggressive missionary to see whether a debater had fully grasped the fullness of his argument. Once Bakhsh brought him the bad news of an alim's conversion to Christianity. Hearing about Moulavi Qudrat Ullah's conversion seemed to have hurt Ahmad deeply. He asked Bakhsh to go back immediately and beg the moulavi to reconsider.

> I'm ready to go there if needed... His name has the word moulavi attached to it. This makes a bad impression on the public... Apostasy from Islam is a big deal. It should not be considered a minor issue. We should bring other people into the fold of Islam. If someone apostates because of our oversight then we are all answerable to God for it... Meet with him alone. Do not meet him in front of others and do not get into a debate because people get defensive. (Ahmad, 1939: 76).

Ahmad argued that Christian missionaries were using every means available to them to convert North Indian Muslims to Christianity and that Muslims response to date had been severely lacking.

> Now O Mussalmans, hear, and hear attentively that this campaign against Islam and Islam's holy influence cannot be defeated by ordinary methods and means. The campaign is backed by the most complex fabrications and the most carefully planned devices that Christians can design. It is pursued mercilessly, with no thought of the expenditure involved. It does not exclude some most shameful devices which decency forbids us to detail. It is a campaign let loose on a large scale and pushed by the worst possible wizardry by Christian peoples, followers of Trinity. Such a campaign cannot be countered by ordinary methods. It cannot be beaten unless its magic and wizardry are confronted by the miraculous power of the Hand of God. Only miraculous power

> can and will smash this campaign and without it we cannot save our simple souls from this vicious Western magic. (Ahmad, 1891b: 5-6)

The ultimate Christian missionary objective, he argues, is the complete destruction of Islam.

> Christians are out to destroy Islam and ready to use lies and fabrications in ways most subtle, and on all occasions and with the help of ever new techniques - all directed to beguile and lead people astray. Christians are defaming and lampooning the Holy Prophet, the perfect man who proved himself the pride of holy men of all times and chief among the saints and apostles of the world. They hesitate not to caricature him in theatrical shows. They try and project a most hateful image they can invent of him. The worst that vicious and unchaste minds can think of is leveled against Islam and the Holy Prophet of Islam to lower them in the eyes of the world. (Ahmad, 1891b: 4-5)

Christian missionary efforts seem to have concerned Ahmad perhaps equally – if not far more than – as the vigorous Hindu Arya efforts that were also aimed at converting Muslims. Basharat Ahmad says that "Ahmad's pen was especially sharp against Christianity" (Ahmad, 1939: 76). In the Muslim attitude towards Hindus one sees hints of Christian missionarie's perceptions of Hindus as primitive heathens who would be easy prey for far more sophisticated Christianity. Another reason for the difference could be the relative status of the two religions in traditional Islamic view. Traditionally, most Muslims regarded Christianity as a religion of divine origin that was founded by one of Allah's major prophets, Hazrat Isa (alt. Maseeh) (Jesus), may peace be on him, who was given the holy book of *Injil*. Hinduism, perceived as idol worship, was considered akin to the polytheistic Quraish of pre-Islamic Arabia. Respect for all of Allah's prophets including Jesus (as indicated by compulsory postfix *allaih-salam* (عليه السلام), meaning 'may peace be on him', following his name) is one of the basic five tenets (pillars) of Islam. This meant that while responding to Arya Hindus a Muslim debater had a freer hand in ridiculing his opponent's arguments and their sources, they would not have been able to do so when responding to a Christian attack. Thus while a Hindu opponent who impugned prophet Muhammad's character could be reminded of Krishna's character flaws, a Christian pursuing the same argument could not be told about any flaws in Jesus' character. Ahmad described the dilemma faced by Muslim debaters in his book *Kitab-ul-Baria*.

> What Christian missionaries want is to be able to disrespect, curse, and blame our prophet peace be upon him as much as they want and hurt us through every humiliation possible but want us to keep our mouth shut and not respond to their attack at all...
>
> Since we accept Hazrat Isa (peace be upon him) to be God's true, pious, and forthright prophet, how can we say any stern words against him but since Christian missionaries do not believe in our prophet, peace be upon him, therefore, they say whatever they want...

There are many missionaries in British India whose job it is to curse our Holy Prophet and lord and master Muhammad peace be upon him, day and night. The worst among them is Rev. Imad-ud-Din Amritsari. In his various books, Tahqiq-ul-Iman etc, he openly lobs obscenities at Muhammad, peace be upon him. He declares him (Muhammad) to be a fraud, usurper of other's women and uses stern and offensive words against him. So does Rev. Thakur Das. In Seerat-ul-Masih and Review Braheen-e-Ahmadiyya, he declares our prophet peace be upon him to be a licentious, a womanizer, a cheat, a thief, a trickster, an ignorant person, and a fraud. In the magazine Dafi-al Buhtan, Rev. Roncallian has used the following words about our prophet: he was licentious, Muhammad's disciples were rapists, swindlers, and thieves. Rev. Rogers writes in his Tafteesh-ul-Islam that Muhammad was licentious, under the thumb of his evil desires, a lecher, a swindler, bloodthirsty, and a liar. In the magazine Nabi Masoon published by American Tract Society, it is written that Muhammad is a sinner, who was guilty of forbidden love meaning rape, and was a swindler and a charlatan. In the magazine Masih-ud-Dajjal, Master Ram Chandar says about our prophet: he was head of dacoits and a thief and a swindler, a womanizer, a liar, licentious, bloodthirsty, and a rapist. In the book "Sawanih Umari Muhammad Sahib," the author, Washington Irwing Sahib writes that Muhammad's disciples were all bandits and swindlers and he himself was self-worshipper, a liar, and a fraudster. In Androona Bible by author Athim Isai, it is written that Muhamad was an anti-Christ and a fraud. Then he says that end for Muhammadans is tragic meaning they will soon disappear. In an issue of Noor Afshan Ludhiana, it is written that Muhammad received Satanic revelations, engaged in improper activities, and was a selfish man, a lost man, a cheater, a fraud, a rapist, a thief, bloodthirsty, a swindler, a dacoit, a helper of Satan, and lustful for his daughter Fatima.

Now it is worth thinking that what can be the consequences of such words against our Holy Prophet peace be upon him. Can a Mussalman ever utter such words against Hazrat Isa allaih-salam?...

These are the stern, liable and denigrating words that missionaries and Arya people have used against our Lord and Master, the Head of prophets, and the Seal of Apostles, peace be upon him. Many of these books have been printed many times in Punjab and Hindustan. They are always assigned to the mission school students to study. They are read in the streets and bazars. Christian women who are assigned to preach take them into Muslim homes. I cannot describe the pain and hurt and disgust with which my trembling body has copied the above words...

What aggressiveness can we show against the missionaries? Because similar to them it is also our duty to accept the dignity and status of Hazrat Isa allaih-salam. Other than the status of God, we consider Isa allaih-salam to be truthful, forthright and deserving of all respect due to a true prophet. But missionaries do not have such good impression of our prophet. (Ahmad, 1898: 93-125)

Ahmad believed that Christians were converting 'thousands upon thousands' of Punjabi Muslims through their clever tactics and needed to be stopped before they succeed in eliminating Islam from the soil of Punjab (M. G. Ahmad, 1891c: 32). Obsessed with the urgent need to find an appropriate response to the Christian onslaught, Ahmad seems to have spent much time praying, fasting, and reading alone in the mosque owned by his family. He also continued to meet like-minded friends to discuss how to respond to Christian missionaries. He seems to have convinced himself that Islamic reforms being advocated by Ahl-e-Hadith, while absolutely necessary, were not enough on their own to defeat Christians. It took him years of prayer, reading, and reflection but by the early 1880s, Ahmad seemed to have discovered a brilliant strategy to free the hands of Muslim debaters. Let's hear Ahmad describe this strategy in his own words.

> I want to express to the readers that my belief in Hazrat Maseeh allaih-salam is a virtuous belief. I believe from the bottom of my heart that he was Allah's true and beloved prophet... Therefore, I grant him all sorts of respect due to his status. But the Yesu that Christians have presented who claimed to be God and considered everyone except himself, all his predecessors to be cursed. That is he considered them to be guilty of such moral deficiencies that deserve to be punished by damnation. I consider such a person to be undeserving of divine blessings. Quran has not informed us of this disrespectful and foul-mouthed Yesu. We are astounded at the manners of this person who permits God to die and claims to be God. And curses such pure souls who were thousands of times better than him. Therefore, in my discussions everywhere, I have referred to this fictional Yesu of Christians. God's humble man Isa bin Mariam (Jesus son of Mary) is absolutely not the target of my crude words. I have had to adopt this strategy after listening to invectives of Christian missionaries for forty years... I have a stack of books by missionaries who have filled their text with hundreds of invectives... It is Christians who are attacking their own Yesu. They just will not stop their cursing and damnation. I am just tired of hearing it. (Ahmad, 2008: 179-180)

Similar to the research efforts undertaken by Dr. Wazir Khan and Moulana Kairanawi in preparation for their famous debate against Pfander, Ahmad then scoured the Christian literature as well as secular-Western critics of Christianity to find evidence that showed Jesus in poor light especially when compared to Prophet Muhammad. He used the evidence he collected in his debates with missionaries that were carried on orally as well as in back-and-forth letters. In a letter to Rev. Soft, he wrote:

> I am surprised that despite all the wretchedness that is found in your beliefs, you people claim that your beliefs are in accordance with reason. I am surprised that people who believe that God gave up his ancient and unchanging majesty to enter into a woman and was birthed through an impure path. Then he bore pain and suffering and died by hanging. (Ahmad, 2008: 103)

In a letter to Rev. Fatah Masih, he wrote:

> The respect that Jesus had among kings of his time is not hidden from you. Those pages are probably still in the Bible where it is written that Herod charged Hazrat Maseeh like criminals towards Pilate. His Godliness did not amount to anything... Was this Godliness?... If you have any honour or shame so then for Maseeh's sake, present this level of respect from kings of the time and get a thousand rupees cash from me... What was Maseeh's mannerism? A glutton and a drunkard, neither pious nor a worshipper, never loved the truth, arrogant, navel-gazer, claimant of God... Disciples ate stolen food in front of your Maseeh. That is breaking chaff from a stranger's fields. Was it the right thing to do?... Maseeh called Peter a Satan. Then why did he forget it and accept Satan into his disciples?... It is a major objection that because Hazrat Maseeh Alaaih-salam was not lucky enough to be endowed with the highest of male virtues and was therefore not able to show a model of true and full domestic behavior with wives. This is why European women have taken advantage of shameful freedom and crossed the limits of moderation and the situation has

> reached such chaos and wretchedness that it cannot be talked about... Maseeh, because of his faulty teachings, left this shortcoming in his sayings and deeds. But because it was a natural urge therefore Europe and Christianity invented rules for it. Now you can see with justice that filthy, dark, morally deficient behavior, and turning of the whole country into a whore-house, the lying on each other of thousands upon thousands like dogs and bitches in Hyde Park during the broad day light... What is the cause of this? This is the result of definitely deficient, unworthy book, anti-nature Paulian Gospel, and half-baked education... I am sending you this letter as a notice that if you keep using such impure words about the Holy Prophet peace be upon him and libel him then I will also take to task your fictional and fake God such that all his Godliness will fall in the pit of humiliation... Surely, whatever you say about the Holy Prophet will also be said of your fictional Maseeh. But we know and believe that true sacred Messiah to be honourable and pure who neither claimed to be a God nor son of God and prophesized the coming of Muhammad, peace and blessings be upon him. (Ahmad, 2008: 187-194)

In a follow up letter to Rev. Masih, he wrote:

> Your fifth objection is that once Muhammad (peace be upon him) happened to look at some woman. When he came home, he had sex with his wife Sauda... You will remember that your God couldn't even restrain himself from the pain of a little hunger and ran towards a fig tree. So can you prove that the tree belonged to his father? Thus the person who couldn't even control himself on seeing someone else's tree and ran to fill his stomach is not only not God but is not even a complete person according to you... Your God didn't get the ability to distinguish but to satisfy the lust of his belly, he ran to a fig tree and didn't even think who it belongs to.
>
> It is strange that an alcoholic, glutton, and drunkard can't be called lascivious but that pure soul whose every action was for God, is allowed to be called licentious by profane people of this world... What can we say and write about your Yesu and how long are we to cry about his behavior? Was it appropriate for him to provide the opportunity for a whore to sit with him at the height of her youth and beauty and with her head bare? And rub her hair on his feet in a flirty and coquettish manner. And massage her perfume of depravity on his head. If Yesu's heart had been free of filthy thoughts, he would have forbidden her from coming near him. But those people who derive pleasure from being touched by depraved women do not listen to any advisor midst such selfish moments. Listen, an honorable dignified man tried to stop Yesu by advising that such actions are not appropriate but Yesu, having understood from his harsh tone that this man is revolted by my deeds, spurned his objection and claimed that this prostitute is very sincere and even you don't have her sincerity. Praise be to Allah. What a fine answer this is! Yesu is praising a harlot that she is very virtuous. Claims of being a God and actions like this! How can we expect piety and virtue from a person who is always inebriated with alcohol and has relations with prostitutes and eats so much that he's known as a glutton? Look at the piety of our lord and master Muhammad, head of all the prophets, may God have peace on him, that he would not even shake hands with women who were pious and virtuous and came to pledge their allegiance to him. Instead he would sit them at a distance and orally teach them to seek forgiveness. But what intelligent and pious person would consider such a person, who doesn't avoid touching young women, as pure of heart? A prostitute is sitting so close as if to be in his arms. Sometimes she stretches her hand to massage his hair and at other times she rubs his feet. Sometime she puts her alluring black hair on his feet. She's doing this show in his lap and Yesu is sitting in a trance. And when someone objects, he scolds them. And situation is this that he is a young alcoholic man who is a bachelor and a pretty young woman is lying in front of him, rubbing her body to his. Is this what gentlemen do? And how can you prove that Yesu's lust did not get stoked

> by her touching? Alas, Yesu did not even have the opportunity to seek company of his wife after looking at this harlot? Who knows what bodily emotions would have been stoked by the touching of that whore and her flirting and coquetting. Lust must have been fully aroused. That is why Yesu couldn't even utter, "O depraved woman, stay away from me." And Injil proves it that that woman was a prostitute and was famous in the whole city for her whoring...
>
> The seventh objection is *muta* (i.e., temporary marriage) being allowed and then disallowed... It is strange as to why Christians mention temporary marriage which is marriage for a fixed time. Why don't they look at their Yesu's behavior? He looks at those young women who he should not have looked at? Was it acceptable for him to sit together with a women? Only if he had been restricted to muta, he could have been saved from such deeds. Did Yesu's old paternal and maternal grandmothers perform muta or did they engage in clear prostitution? (Ahmad, 2008: 223-228)

In a letter to a Barelvi Muslims, he wrote:

> These people [Christians] do not look at their own scriptures and see how open Gospels are to objections. Listen, how great of an objection is this that Mary was given over to Jerusalem Temple to serve there her entire life and never get a husband. But when her pregnancy became visible after six or seven months, then within the condition of pregnancy, she was married of to a carpenter named Joseph. One or two months after her moving to his house, a son was born to Mary who was named Isa or Yesu. Now the objection is that if the pregnancy was indeed miraculous then why didn't they wait till its completion? The second objection is that the covenant was that Mary would serve at the temple for a long time then why was the covenant broken and she made the wife of Joseph the carpenter? The third objection is that since it was proscribed by Torah to marry a pregnant woman, why was Mary married to Joseph in contravention of the orders of Torah? Even though Joseph had a previous wife and he was angered by this marriage. It seems that those who deny multiple marriages have not been informed about this marriage. Thus a detractor would be justified to assume that the reason for this marriage was that elders of the nation had suspected her of having an inappropriate pregnancy. Even though we believe according to the Holy Quran that the pregnancy was entirely through God's power so that it is a sign for the Jews till the end of times. And the same way that thousands of bugs are born by themselves during the rainy season, Hazrat Adam was also born without a father or a mother. So then a fatherless birth for Hazrat Isa does not prove a high status. Actually, being born without a father indicates a lack of certain qualities. In summary, Hazrat Mary's betrothal only happened because of a suspicion, otherwise a why would a woman who had been anointed to serve the Jerusalem Temple need a marriage? (Ahmad, 2008: 288-289)

There was little mention of Ahmad's new strategy of distinguishing the Christian Jesus from Muslim Jesus in Ahmad's first book titled Braheen-e-Ahmadiyya that was completed in 1879 but took another five years to publish because of lack of funds (Dard, 1948: 90). This could also be because Braheen was primarily directed at proving that Islam was superior to Hinduism to respond to Arya Samaj's missionary activities. In Braheen, Ahmad displayed the same fierceness in attacking Hinduism as he was to go on to display in attacking Christianity. Publication of Braheen also occurred at a time when Muslims were still fearful of appearing too aggressive lest they be labeled as *Wahabi* extremists by the British government. In fact, Muhammad

Hussain Batalavi actively lobbied the British officials throughout the 1880s requesting that reformist Muslims from Shah-Waliullah's tradition be relabeled an Ahl-e-Hadith rather than as Wahabis.

Ahmad shared drafts of Braheen with Batalavi to seek his comments. Batalavi published advertisements in Ishat-us-Sunnah promoting Braheen years before its publications. Finally, when the book was published, Batalavi and most in the Ahl-e-Hadith community (especially those in Delhi and Punjab) hailed it enthusiastically. Batalavi wrote a glowing review in Ishat-us-Sunnah.

> In our opinion, it is in this time and in the present circumstances, a book the like of which has not been written up to this time in Islam, and nothing can be said about the future; Allah may bring about another affair after this. Its author, too, has proved himself firm in helping the cause of Islam, with his property, his person, his pen, his tongue and his personal religious experience, to such an extent that it is rarely seen among Muslims who have gone before. If someone thinks that my words are Asian exaggeration then show me at least one such book that confronts opponents of Islam especially the Arya Samaj with such gusto and enthusiasm. And point out such supporters of Islam who have taken upon themselves to help the cause of Islam with their property, their person, their treasure, their pen, and their tongue. And who has successfully challenged, with all his manly courage, opponents of Islam and deniers of revelation that if they doubt divine revelation to visit him and experience, observe and taste it. (Batalavi, 1884: 169-170)

The reaction to Braheen was not uniform even among the Ahl-e-Hadith. Nawab Sidiq Hasan Khan, for instance, angrily returned copies of Ahmad's books sent to him with a note saying that he was afraid that such controversial works will draw the anger of the British authorities. Responding to Khan in an *ishtihar* (i.e., poster), Ahmad exhibited a much more sophisticated and nuanced understanding of the relationship between the British government and Christian missionaries in India than was common among Muslim leaders of the time. He wrote:

> Criticism of the government in such matters is wrong. It is not the British government's policy to prevent a nation from proving the truth of its religion or prevent people from supporting religious books. Yes, if there is something that obstructs peace or is against administration of the state then the government will interfere in it. Otherwise, everyone has the permission to use resources for the progress of their religion... This is the reason that about a month ago Lieutenant Government of our country of Punjab, Sir Charles Aitcheson Sahib Bahadur, came to the district of Gurdaspur. While laying down the foundations of a church, with remarkable plainness and without a hint of pretense, he said that, "I hope that within a few days this country will progress in religiosity and forthrightness"... One should understand the freedom of the British government from this remark. Because when the Lieutenant Governor Bahadur himself not only wants to spread his good faith in India but also urges others whenever he finds an occasion then why would he be angry when others express their affinity for their religion? (Ahmad, 1891c: 53)

Even among the pro-British Ahl-e-Hadith, Ahmad's unreserved support for all British government actions stood out and marked him along with Sir Syed Ahmad Khan as one of the most pro-government Muslim leaders of his time.

While Nawab Sidiq Khan had criticized Braheen because of its anti-Christian views, a number of other Muslim leaders saw in it claims that ran counter to their understanding of Islamic doctrine. These included Amritsar and Ludhiana's Muslim leaders who issued a fatwa of kufr against Ahmad and went around India to collect signatures from other Muslims to the same effect (Batalavi, 1884: 170). Ahmad's friend, Muhammad Hussain Batalavi, took upon himself to reply to this criticism through his *Ishat-us-Sunnah* magazine. He devoted a whopping one hundred and fifty three pages of Issue Nine and Ten to respond to each of these criticisms with well researched arguments. The amount of time, effort, money, and his personal capital, Batalavi spent on defending Ahmad reveals the extent of their ideological and social connections. The review also reveals Batalavi's perceptions of Ahmad. Batalavi clearly identifies with Ahmad as a fellow Punjabi Ahl-e-Hadith who is willing to spend considerable resources to defend Islam. He also sees Ahmad as more rural, less educated, less sophisticated, and less successful than himself. He sees Ahmad as "a Punjabi who has never had the opportunity" to live in the cosmopolitan cultural centers of Hindustan, "who hasn't had the occasion to read Urdu literature" and therefore is not able to write "refined Urdu vernacular" (Batalavi, 1884: 346). There are also hints of a noble savage in Batalavi's perceptions of Ahmad as he sees Ahmad as someone who is so overcome with religious fervor and zeal that "he's unable to hold back" from including his unrelated revelations in the Braheen (thereby lengthening it and "increasing publication costs"), and someone who doesn't understand that the "current civilization" demands refraining from crudely attacking one's enemies (Batalavi, 1884: 346).

Batalavi touts his personal knowledge of Ahmad's beliefs beyond the words written in Braheen to defend Ahmad. Answering the accusation that some of Ahmad's English revelations are grammatically incorrect, Batalavi says, "When I met the author who visited the city of Batala, where I am now, I asked him, 'when you receive revelations in English, are you shown English alphabets or Persian alphabets?' He responded that he is shown English sentences written in Farsi script. That's when I became sure of my suggestion that the mistake lies in the author's perception... and not the divine revelation" (Batalavi, 1884: 291). Ahmad's ignorance of English and his miracle of English revelations will attract English speaking Christians and Hindus to Islam argues Batalavi. It's clear that Batalavi sees the younger Ahmad as his junior, albeit more zealous, friend in need of assistance. Batalavi sees himself as heroically defending Ahmad with his superior knowledge of Quran, Hadith, and Islamic traditions. Since Batalavi does not see Ahmad as his competitor, he holds nothing back in defending him.

> Perhaps some of our critics will place me in the same category as the author of Braheen-e-Ahmadiyya and slap the fatwa of kufr on me as well. They may say that I have raised the author of Braheen to the level of prophet Muhammad and I have declared his revelations to be innocent similar to the revelations of the Prophet but I am not afraid of their fatwa of kufr... (Batalavi, 1884: 284)

Over and over again, Batalavi insists throughout his review that Mirza Ghulam Ahmad had not made any prophetic claims (pages 175; 191; 260; 268; 269; 273; 275; 278; 279). Muslim leaders who insist that he has, are mistaken argues Batalavi. Mirza Ghulam Ahmad is a Muslim and not a *kafir*, argues Batalavi. In one paragraph where Batalavi not only foresees the group-name of *Ahmadi* that Ahmad was to assign to his movement seventeen years later but also the pejorative group name of *Mirzai* that their opponents were to call them by, Batalavi writes:

> What does he [Mirza Ghulam Ahmad] conclude from his revelations and miracles? Does he use them to prove his prophethood or the prophethood of Muhammad? What religion does he invite people (including many top padres, pundits, Brahmo Arya rajas, and sardars of other religions) with such bravery and boldness? Is it the Islamic religion or Ahmadi religion or Mirzai religion?
>
> Unless you are a person whose heart has been darkened with prejudice, these arguments and reasoning would have convinced you that **he absolutely makes no claim whatsoever to his own prophethood** [emphasis in the original]. The true purpose of all of each and every one of his claims is the proof of prophethood of Muhammad. (Batalavi, 1884: 278-279)

Thanks to the tireless promotion of reformist Muslims leaders such as Batalavi, Ahmad started to become known all across North India as a fierce advocate of the Islamic causes. While Ahmad was having some success in his professional life, his personal life was going through some turmoil. His elder brother, who had been his caretaker since his father's death, also died in 1883. Ahmad's grown up sons had their own lives (the older son, Sultan Ahmad, had moved away from Qadian to work as a civil servant). Ahmad's relations with his first wife were 'practically severed' perhaps because of her insistence to cling to her traditional orthodox Hanafi beliefs and because of Ahmad now openly associating himself with his unorthodox Ahl-e-Hadith friends such as Batalavi. Ahmad felt the need for companionship and decided that the time was right to get married again but this time to an Ahl-e-Hadith girl who shared his reformist views. According to the account, written by Ahmad's son, he contacted his friend Muhammad Hussain Batalavi requesting him to look for a suitable match for him.

> My mother told me that, "before my marriage, Hazrat Sahib [i.e., Ahmad] found out that his second marriage would be in Delhi. Therefore, he mentioned it to Moulavi Muhammad Hussain Batalavi. Because at that time he [i.e., Batalavi] had a list of all Ahl-e-Hadith girls and Mir Sahib [i.e., her father and Ahmad's future father in law] was also an Ahl-e-Hadith who had a great relationship with him. Therefore, he mentioned Hazrat Sahib's name to Mir Sahib. He also wrote to Mir Sahib. In the beginning, Mir Sahib did not like this match because of the differences in our ages but eventually he relented. Then Hazrat Sahib came to Delhi to marry me. He was accompanied by Shaikh Hamid Ali and Lala Malawa Mal. The marriage was performed by Moulana Nazir Hussain. This happened on Monday, 27 Muharram 1302, when I was eighteen years old. Hazrat Sahib gave Moulavi Nazir Hussain five rupees and a prayer rug as a nazar. (Ahmad, 1935: 57-58)

Offering *nazar* to the founder of the Ahl-e-Hadith movement in India is a rare occasion where Ahmad's Ahmadi biographers allow him to engage in a traditional Sufi practice. It is noteworthy that this exception was made for an Ahl-e-Hadith alim and not for a traditional Sufi. It appears that Nazir Hussain shared the affection that Ahmad had for him. According to Ahmad's disciple and biographer Pir Siraj-ul-Haq Numani, Nazir Hussain called Ahmad a 'holy man' while Ahmad called Nazir a '*waliullah*'.

> One day I asked Hazrat Aqdas[as] [i.e., Mirza Ghulam Ahmad] "Hazoor, what do you say about the *wahabi ghair-muqalads*? Are they an impure sect? (I used to be a strongly prejudiced Hanafi back then)."
>
> On listening to this, he smiled but stayed silent and didn't say a word.
>
> Next day, when I asked him again he said, "that sect is also from God and is not bad. When people stressed taqlid and Hanafism so much that they turned the four saints into prophets then God created this sect with his wisdom so that muqallids return to the right path and become balanced. The only thing bad about them is that each of them has become a mujtahid and an imam and started calling the four imams bad."
>
> I asked, "what do you think about the founder and the leader of this sect, Moulavi Nazir Hussain?"
>
> He said, "I have good impression of Moulavi Nazir Hussain and he is also a *waliullah*."
>
> I said that "Huzoor, I should ask for forgiveness for many bad things that I have uttered about Nazir Hussain"...
>
> I traveled to Qadian, I said I am coming from Qadian. He asked, "did you also meet Mirza Ghulam Ahmad? " I said, "Yes I just came from his place." He said, "It's good that you went there and met him. He is a holy person similar to the holy people of the past"... He said, "if you go to Qadian, then convey my salam greetings in accordance with the Sunnah of the holy prophet." (Numani, 1915: 214-215)

Walliullah literally means friend of Allah and is considered to be an extremely high rank on the ladder of piousness. Ahmad's use of the title to refer to Syed Nazeer Hussain show his reverence for the Ahl-e-Hadith leader.

While Ahmad's innovative strategy of aggressively attacking Christians by distinguishing the Christian Jesus from Muslims' Issa won him accolades from fellow Muslims, it failed to stem the tide of Muslim conversions to Christianity. After about a decade (from about 1880 to about 1890) of exclusively relying on this strategy, Ahmad came to the conclusion that the strategy was not enough on its own. As the number of Christian missions, schools, and hospitals in Punjab multiplied, the number of Muslims converting to Christianity also continued to increase. In fact, some of Ahmad's followers and relatives also converted to Christianity (Ahmad, 1898: 143). Ahmad searched for an additional weapon to counter the attack of Christian missionaries against Islam. This is when he discovered what he regarded as his greatest weapon.

Ahmad came to call it the key to "breaking the cross" i.e., debunking Christianity and banishing it from earth forever.

> At long last, the time for the breaking of the cross came: the time when the error of the creed of the cross was to be exposed with such clarity as the splitting a piece of wood into two. Heaven has thrown open the door to the breaking of the cross, so that whoever seeks the truth may seek and learn. (Ahmad, 1908: 72)

In 1891, Ahmad published three books titled "Fath-e-Islam," "Tauziay Maram," and "Azala-Auoham" to outline his new strategy. He claimed that God had told him that (a) Jesus had not been lifted to heaven by God prior to his crucifixion, (b) it had been Jesus who had suffered the indignity of having been hanged on a cross but God has miraculously saved him from immediate death, and (c) Jesus had died a natural death after having lived his full natural life.

In order to sell this change in belief to Muslims, Ahmad adopted an arcing narrative that appealed to social identity beliefs of his fellow Muslims (Upal, 2015c). He argued that Muslims during their glory days (i.e., during the life of the prophet and his first four 'rightly guided' Caliphs) did not believe in Jesus' physical ascension to heaven. Such beliefs he argued were corruptions that had slowly crept into Islam over the preceding centuries. Note that this is the same process through which Muslims believe that the divine message to Christians and Jews was corrupted over time. Ahmad merely extended the same process to the Islamic beliefs regarding Jesus. He argued that a correction in Islamic beliefs was the only way for Muslims to fight a well-organized and well-funded Christian missionary effort. If Muslims adopt these changes, Ahmad argued that they will be able to turn the world back to the glory days when Islam dominated Christianity and this would result in the long-promised final victory of Islam. In order to appeal to Muslim social identity beliefs, he argued that a natural death for Jesus would show that Jesus was not better than Muhammad. Thus, those who oppose this change are not true friends of Islam.

> To believe that Jesus is alive, is highly insulting and derogatory to the Holy Prophet. I cannot stand this sacrilege even for a moment. Everyone knows that the Holy Prophet passed away at the age of sixty-three and lies buried in his tomb at Medina, which millions of pilgrims visit every year. If it is disrespectful to believe in the death of Jesus or even to think of it, then I ask how can you permit this insolence and disrespect with regard to the Holy Prophet?... how can one claim to love and be a follower of the Holy Prophet if he accepts a superior status for Jesus by pronouncing him alive and the Holy Prophet dead?
>
> Believe me, had the Holy Prophet continued to live, not a single person would have remained an unbeliever. On the other hand, what is the use of Jesus supposedly being alive except for the four hundred million Christians? Just pause and ponder! Haven't you tested the belief in Jesus being alive? Hasn't the outcome been grave? Name any Muslim community of which no member has been converted to Christianity. For my part, I can say with certainty that Muslims belonging to all classes have converted to Christianity and the number of such converts may be more than one hundred thousand. The lone weapon with which Christians convert Muslims to Christianity

> is none other than the belief that Jesus is still alive. "Show us such an exclusive quality in anyone else other than Jesus," they say. "If Jesus is not God, why was he singled out for this distinction?" To them, he is Ever-Living, and Self-Existing—God forbid. This concept of a living Jesus has emboldened them to mount attack on Muslims, the effect of which I have already pointed out. Imagine the dismay of the priests if, on the contrary, you were to establish the death of Jesus! I have confronted many an eminent Christian priest with this and they have invariably replied: "If it is proved that Jesus is dead, our faith cannot live."
>
> Here is another point to ponder: You have seen for yourselves the consequences of the belief in a living Jesus. Now try to visualize the after-effect of his demise and the damage it does to the Christian creed. This can be illustrated from the fact that whenever any of my followers wants to enter into discussion with the Christians on this point, they instantly refuse, for they know it would sound their death knell. With Jesus dead, they can prove neither redemption, nor Jesus' divinity, nor, for that matter, his being the son of God. So try this formula for a few days and the reality will speak for itself...this too is an undeniable fact that Islam lives if Jesus dies. If you reflect on this, you will realize that this is the issue which is destined to seal the fate of Christianity. It is the mainstay of the edifice of the Christian faith, so let it collapse. (Ahmad, 1908: 72)

While Ahmad had hoped to eventually turn the focus of his missionary efforts towards the enemy camp and convert Christians to Islam, the primary target audience of his early messages consisted of his fellow Ahl-e-Hadith including Muhammad Hussain Batalavi, Syed Nazir Hussain, and sons of late Moulavi Abdullah Ghaznavi. In his 1891 books, in a traditional Ahl-e-Hadith style, he bypassed the medieval Islamic sources of authority such as the four Imams and focused directly on evidence from Quran and Hadith to convince his audience of the merits of his ijtehad.

To really blunt the arguments by Christian Evangelists, who pointed to Muhammad's tomb as a proof of his lower status, Ahmad wanted a physical symbol of Jesus' death, preferably a tomb. Ahmad's search for Jesus' tomb seemed to have started in the holy land where Christian sources have traditionally placed Jesus' tomb. In 1891, he wrote that, "Of course it is true that Jesus died in Galilee but it is not true that his body was resurrected" (Ahmad, 1891a: 473). In 1894, he wrote to a Syrian acquaintance inquiring about the exact co-ordinates of the tomb. When told that it was nearby, he assumed that it was in Syria. He wrote,

> ...the funny thing is that there is a tomb of Jesus in the country of Syria. For further clarity regarding this matter I quote the witness of brother Syed Muhammad Al-saeedi Tarablassi who lives in Tarablas, Syria. . . If you were to argue that the tomb is fake then you would have to prove your argument. You would also have to show when the fakery were invented? If Jesus' tomb is proved fake we would also become suspicious about the tombs of other prophets and lose our belief in their authenticity. We would have to admit that perhaps those tombs are also fake. (Ahmad, 1894: 18-19)

The milieu in which Ahmad made his claims served as fodder for his creative process both by providing him with social problems to think about and by limiting the creative space he had to explore to find solutions. The period in which Ahmad lived was unique in a number of ways: opening of reliable contacts with India and Americas unleashed an exciting period of discovery for Europeans, contacts with India allowed them to explore the common origins of the European and Indian languages (Latin, Greek, and Sanskrit), and European and Indian religions (Christianity and Budhism). Some were reminded of the holy land as described in the Bible when they visited India. Some saw the lighter skinned Afghans and Kashmiris as being similar to European Aryans while others saw them as being similar to Jews in appearance as well as customs. The presumed similarities between Buddhism and Christianity also led scholars to a number of controversial hypotheses. Perhaps, none was as controversial as the hypothesis put forward by Russian journalist Nicolas Notovitch, who speculated that Jesus had come to India during his "missing years" (i.e., between the ages of 12 and 30). He claimed to have personally seen the evidence in a Buddhist book shown to him by monks when he visited Tibet (Notovitch, 1990/1887).

It was in this milieu that Ahmad first articulated his claim that Jesus had come to India after having miraculously survived the cross to save the lost tribes that lives in Afghanistan and Kashmir. After preaching to all twelve Israeli tribes, Jesus had died a natural death at the ripe old age of 120 and was buried in Srinagar, Kashmir where one of Ahmad's followers had located the tomb of an unknown Christian. Given the time period during which it was proposed, nothing about Ahmad's theory sounds remarkable (for example, when compared with Notovitch's theory) other than the fact that after narrating all the scholarly arguments in favour of his claim, Ahmad concluded that it was God who had revealed the truth to him.

Belief in Allah as the only God and Muhammad as his messenger has always been considered by most Muslims to be the primary condition of being a Muslim. Jesus' being alive or dead had never been the primary determinant of Muslims identity. Since Jesus is not central to the Islamic narrative, it is not surprising if an average Muslim does not have a strong position on the issue of Jesus' life and death. The only reason the issue aroused emotions among nineteenth century North Indian Muslims was that the Christian missionaries had been introducing it to open discussions with Muslims. Thus, had it not been for the Christian missionaries bringing Jesus to the fore of the consciousness of North Indian Muslims, it is doubtful that anyone would have paid attention to Ahmad's message of Jesus' survival of his crucifixion and his later death. Similarly, it is hard to imagine Ahmad's claims gaining much traction in a pre-reform traditional Hanafi/Sufi environment of North Indian Islam. This is because in such an environment, doing taqlid (i.e., the following of more-than-a millennia of tradition) is what makes one a better Muslim and not engaging in ijtehad (i.e., using one's deep knowledge of Quran & Hadith to address current problems). One can also imagine how difficult Ahmad's task would have been if in addition to reinterpreting Quran and Hadith to suggest that Jesus had died, Ahmad also had to reinterpret sayings of hundreds of Hanafi scholars on the issue. However, because similar to his Ahl-e-Hadith friends, he insisted that Quran and authentic Hadith were the only sources of authority, this was not an issue in his mind or his closest associates. While there is no evidence to indicate that Ahmad ever formally offered bayat to an Ahl-e-Hadith alim (such as Syed Nazir Hussain), he regarded them highly in the early part of his life and many of his associates openly identified themselves as Ahl-e-Hadith prior to becoming Ahmadis. Ahmad also called himself a ghair-muqallid and acknowledged that that is how others saw him. In a letter written to Moulavi Nur-ud-Din, he wrote:

> This humble man had sent out many letter for your second marriage. One place that has responded seems in accordance with our desires. By this I refer to Mir Abbas Ali Shah Sahib's letter that I am sending for your perusal. This letter has a strange condition that [groom] be Hanafi, not be a ghair-muqalad. Since Mir Sahib is Hanafi and my sincere friend Munshi Ahmad Jan Sahib (May God shower him with blessings), with whose blessed daughter this proposal is in front of us, was also a staunch Hanafi. That's why they have put the condition of Hanafi. Even though all Muslims are part of Hanafi religion of Abraham but still there has to be an appropriate answer. Munshi Ahmad Jan... was not prejudiced. He knew about me that that I am not a believer in Hanafi taqlid and nor did he like it but even then this thought did not prevent him from love and sincerity. (Ahmad, 2008: 56)

7.1 How Ahmad's Friends & Family Helped Him Realize His Role in Life

One of the people who read Braheen was an Ahl-e-Hadith *alim* by the name of Hakeem Nur-ud-Din. As an Ahl-e-Hadith Muslim ashraf who worked as a chief physician for the Maharaja of Kashmir, Nur-ud-Din was exactly the kind of person Ahmad had hoped his books would appeal to. Nur-ud-Din was a sanad-holding alim who had travelled all over India and Arabia to study Islam under the most well-known ulema of the time. Nur-ud-Din was so impressed with Braheen that he decided to pay a visit to Qadian to meet with its author. They seem to have hit it off from their first meeting in 1885:

> As soon as I saw his face I was overjoyed, and felt happy and grateful to have found the *perfect man* that I was seeking all my life... (Nur-ud-Din, 1908)

Nur-ud-Din said that in that moment he felt such an instantaneous connection with Ahmad that he knew that "I could lay down my life for him."

> At the end of the first meeting, I offered my hand for Bay'ah. Hazrat Mirza Sahib [i.e., Mirza Ghulam Ahmad] said, he was not yet divinely commissioned to accept bay'ah; then I made Mirza Sahib promise me that I would be the person whose bay'ah would be accepted first.
>
> You may be sure.
>
> Is there any special effort that I should put forth to win the love of God?
>
> Write a book on refutation of Christian criticism of Islam. (Nur-ud-Din, 1908)

Nur-ud-Din devoted the next few months to writ the book he was asked to write. Over the next few years, Nur-ud-Din was to completely devote everything he had (his time, money, and expertise) to Ahmad and the two would have such an influence on each other that it is hard to imagine what course Ahmad's life would have taken had he not met Nur-ud-Din. They wrote to each other frequently. Nur-ud-Din visited Qadian as frequently as he could. When Nur-ud-Din got sick, Ahmad visited him in Jammu. Nur-ud-Din kept urging Ahmad to accept a leadership role and to accept his oath of allegiance (bay'ah) but Ahmad was reluctant to do so.

It is hard to know what was going through Ahmad's mind as he repeatedly declined invitations from devotees such as Nur-ud-Din to publicly proclaim divine authority and to accept their formal oaths of allegiance. Ahmad's Muslims opponents often attacked him for lacking the authority of a formal sanad from any well-known alim. They argued that this was a necessary, though not sufficient, condition for engaging in ijtehad. Only a much higher ranked mujaddid could advocate the kind of changes Ahmad was advocating. Islamic tradition describes a hierarchy of divine authorities ranging from 'Christian saint-like' figures called a *mujaddid* (literally a renewer) to prophets called a *nabi* or a *rasul*. A mujaddid introduces reforms in religious doctrines but failure to recognize them and give them one's allegiance (or bayah) does not make a Muslim an infidel. Prophets on the other hand, bring new laws and scriptures from God and have to be obeyed to achieve salvation. Claimants to the offices of mujaddid and nabi/rasul have been treated very differently in the history of Islam. Most claimants to the office of muijaddid (e.g., Shah Waliullah of Delhi) were tolerated by most Muslims and even revered by some. All of those who claimed to be prophets, however, were vehemently opposed and attempts on their life were religiously sanctioned by a vast majority of Muslims. These include a series of tribal chiefs who declared themselves to be prophets following the death of Prophet Muhammad in 632 AD. They were militarily attacked, defeated and killed under the commands of the first successor (or Caliph) to the Prophet Muhammad.

In 1882, Ahmad claimed that God had appointed him a mujaddid of the 14th Islamic century. As he was to explain to Nur-ud-Din years later, as a mujaddid, he was not authorized to accept people's formalallegiance or to start a formal community. Nur-ud-Din's insistence that he accept his bayah may have indicated to Ahmad that Nur-ud-Din thought that Ahmad deserved a higher rank. After desertions of some prominent followers following Ahmad's son Bashir's death, who he had thought was the target of his prophecy of a wildly successful successor, died in infancy, Ahmad decided to formalize his relationship with his followers in 1889. Nur-ud-Din became the first of the approximately 40 people to offer their oath to Ahmad. In 1891, Ahmad formally announced the formation of the Ahmadiyya Jama'at as a chosen community of true Muslims who would win the final global victory of Islam by converting people around the globe to 'true Islam.' He also claimed that God had told him that the Mahdi and the Messiah were supposed to be the same person and that God had appointed him to both offices. He said that Allah had told him "the Messiah, son of Mary, prophet of

Allah, had died and in his attribute thou hast come in accordance with the promise." (Mirza Ghulam Ahmad's 1890 revelation as quoted by Alladin, 2015).

Since the original Jesus was a prophet, Ahmad seemed to suggest in some of his writings that he was also raised to that status so that the Islamic Messiah was seen by people as inferior to the Christian Messiah. He argued that a rise in his personal status was also needed to show that Muhammad was superior to Jesus.

> It pains me to hear the ignorant people say that Jesus is alive in heaven, whereas it is in the Holy Prophet that I see the signs of life. People ask me how could I have claimed to be the Promised Messiah. Let me tell them that, through complete obedience to the Holy Prophet, one can attain a status even higher than that of Jesus… (Ahmad, 2007: 26)

When some Muslims objected that the Quran declares Muhammad as his final prophet (e.g., 33:40 "Muhammad is not the father of any man among you, but he is the Messenger of God and the last of the prophets"), he argued that they had misunderstood the true meaning of the verse. He said that the Arabic word "khatim" is not used in the sense of its common meaning of "last" but in the less used meaning of a "seal."

> the verse means that, though the Holy Prophet did not have any male offspring, he will have countless spiritual progeny, and that he is 'the Seal of Prophets', which means that no one can attain the excellence of prophethood unless he possesses the certificate of obedience to him. This is the true connotation of this verse, but these people have completely reversed its meaning and have rejected the bounty of prophethood in the future, even though this implies a criticism of the Holy Prophet himself. The perfection of a Prophet lies in his ability to impart to his followers the excellence of prophethood… Remember O Muslims! It is the height of ignorance and stupidity to harbour this belief. If Islam is indeed such a 'dead' religion, then whom will you invite to embrace it? Will you take its corpse to Japan or will you offer it to Europe?… The fact, O naive and blind people, is that our Holy Prophet, and our lord and master, (countless blessings be upon him), surpassed all Prophets in his spiritual influence. The influence of all prophets came to an end at a certain point and their people and their religions have no trace of life left in them, but the spiritual influence of the Holy Prophet will endure to the last day. For this reason, this umma does not require that a Messiah should come into it from outside, for under the Holy Prophet's influence, even an ordinary man can become a Messiah, just as God has done in my case. (Ahmad, 1906: 74-75)

By following Muhammad, Ahmad said that he had become not only similar to Jesus but also similar to numerous other prophets who had all become prophets because of Muhammad the *khatim* i.e., prophet-maker in Ahmad's new interpretation. This led him to famously declare that, "I am Adam, I am Noah, I am Issac, I am Jacob, I am Ismail, I am Moses, I am Abraham, I am Jesus, and I am Muhammad" (M. G. Ahmad, 1907: 84).

Ahmad also announced that God had told him that traditional Muslim beliefs about Jihad also needed to be reformed. The abrogation of Jihad as "war against infidels" was also required to ensure the domination of Islam over Christianity, argued Ahmad.

Here's how one of Ahmad's early followers, Moulana Muhammad Ali interpreted Ahmad's claims regarding Jihad:

> What Ahmad rejected was not the doctrine of jihad but the orthodox interpretation of it, which had given rise in the West to grave misconceptions regarding the doctrine of jihad, so that even unprejudiced Western writers thought the word jihad to be synonymous with war undertaken for forcing the religion of Islam upon non-Muslims. Thus, in the Encyclopedia of Islam, the article on Jihad opens with the following words: 'The spread of Islam by arms is a religious duty upon Muslims in general'. Klein, in his Religion of Islam, makes an even more sweeping statement: 'Jihad ... The fighting against unbelievers with the object of either winning them over to Islam, or subduing and exterminating them in case they refuse to become Muslims.'
>
> In the Muslim popular mind there was an even-greater misconception, that the killing of an unbeliever was jihad and that such an act entitled the perpetrator to be called a ghazi.
>
> This conception, coupled with the prevailing belief in the advent of a Mahdi who would put all non-Muslims to the sword if they refused to accept Islam, opposed as it was to the plain teachings of the Holy Quran, was doing immense harm to the cause of the spread of Islam among non-Muslims. (Ali, 1937: 75)

While Ahmad's message quickly attracted dozens of sympathetic souls from among the Ahl-e-Hadith who already believed that Jihad against the British was forbidden, he also attracted plenty of opposition, not only from Christians and Hindus but also fellow Muslims, perhaps most shocking even from fellow Ahl-e-Hadith, i.e. Muhammad Hussain Batalavi, who in a complete role reversal now vowed to bring Ahmad down. Batalavi lobbied his teacher Syed Nazir Hussain to also publicly condemn Ahmad as *kafir*. According to Ahmadi sources, the old and frail Hussain only begrudgingly agreed to Batalavi's repeated requests. Abdullah Ghaznavi's sons (who succeeded him in Khjayrdi) who had never liked their father's close relationship with Ahmad, also joined in condemning him as a *kafir*. Ahmad described some of the fatwas issued against him by his former Ahl-e-Hadith supporters.

> **Mian Nazeer Hussain Dehlavi known as Shaikh-ul-kul**
>
> He is the writer and originator of the fatwa that was published against me in Ishat-us-Sunnah Volume 5 Issue 13. The writer of the fatwa i.e., Mian Sahib has used these words about me in his fatwa:
>
> Expelled from Ahl-e-Sunnat. His actual practice is that of inner atheists etc. who have become lost. He can be called one of the thirty dajjals (anti-Christs) that have been foretold in the Hadith because of his false claims and publication and non-believing ways...
>
> **Shaikh Muhammad Hussain Batalavi, Editor of Ishat-us-Sunnah...**
>
> A hidden enemy of Islam, the second Musailima, dajjal of the age, fortune-teller, astrologer...

Ghaznavi Group

Moulavi Abdul Jabbar Sahib while signing the above mentioned fatwa on page 200 wrote the following words: "The one who makes such claims is against the Prophet of God... One of those people about whom Allah's prophet has said that at the end of times lying dajjals will be born... Save yourself from them lest they misguide you and talk you into it. His (Qadiani's) little chicks (followers) are eunuchs of Christianity and Hinduism."...

Abdul haq Ghaznavi...

Dajjal, atheist, liar, black-hearted, evil-doer, Satan, cursed, faithless, dishonored... (Ahmad, 1898: 118-121)

Given the fierce opposition from his fellow Ahl-e-hadith, Ahmad seemed to have backed off a little from his original prophetic claims. In a pamphlet Ahmad published on October 2, 1891, he wrote:

I have heard that some leading ulema of this city of Delhi give publicity to this charge against me that I lay claim to prophethood... Therefore, to make manifest the truth I declare before the learned and the laymen that this charge is a fabrication, pure and simple. Neither, do I lay claim to prophethood, nor am I a denier of miracles and angels... After our leader and master Muhammad, peace be upon him, the last of the messengers, I consider anyone who claims prophethood and messengership to be a liar and kafir. (Ahmad, 1891c: 230)

On 3rd February 1892, he wrote:

Hazrat Muhammad Mustafa, peace and the blessings of God be upon him, is the last of the Prophets. I, therefore, wish to make it clear to all my Muslims brothers that if these words vex or wound their feelings in any ways, they may look upon them as having been struck off and replaced by the word Muhaddath everywhere... it has never been my intention to take the word 'Nabi' to suggest Real Prophethood. The word, to put in a plain language, means a Muhaddath which the Holy Prophet has interpreted to denote 'one spoken by God.' (Ali, 1917: 3-4)

In 1897, he wrote:

Can a wretched imposter who claims messengership and prophethood for himself have any belief in the Holy Quran? And can a man who believes in the Holy Quran, and believes the verse `*He is the Messenger of Allah and the Khatam an-nabiyyin'* to be the word of God, say that he is a messenger and prophet after the Holy Prophet Muhammad?

Anyone who is fair-minded should remember that I have never, at any time, made a claim of nubuwwat or risalat [prophethood or messengership] in the real sense. To use a word in a non-real sense, and to employ it in speech according to its broad, root meaning, does not imply infidelity (kufr). However, I do not like even this much, for there is the possibility that ordinary Muslims may misunderstand it. (Ahmad, 1897: 27 Footnote)

In order to change beliefs sanctioned by generations of Muslims for hundreds of years including some of the most revered Islamic figures, Ahmad knew that he still needed the divine authority superseding all Muslims since the Prophet Muhammad's time, but since the Muslim aversion to the word prophet seemed to distract attention from the merits of the reformations he was advocating, he decided to invent a new religious authority. He called it, "zilli nabi" or "shadowy prophethood". He claimed that since Imam Mahdi and Promised Messiah was to be a zill (i.e., shadow) of two prophets (namely Muhammad and Jesus), he could also be called a zilli prophet i.e., his prophethood was not an independent prophethood but only a reflection of the prophethood of Jesus and Muhammad. While this new invention may have allowed him some breathing room, it would also, in time, split his community into two with some Ahmadis arguing that he was a full prophet while others denying it - but let's not get ahead of ourselves. After securing his rear flank, Ahmad focused his attention back onto Christians and Hindus.

Ahmad held a number of public debates with leaders of major religious factions including Dr. Henry Martyn Clark of the CMS. At the end of the debate that lasted fifteen days, Ahmad announced that God had revealed to him that within fifteen months God would show a sign in support of Islam. According to Rev. Robert Clark, this was interpreted by most as indicating that he was to die within fifteen months.

> There is great excitement about the Mullah's prophecy about the death of Athim and of Henry in fifteen months after the great controversy in which they took part. The Easterns we shall never understand. (Clark, 1907)

This was not the first time Ahmad had been understood to have prophesized the death of an opponent. In 1893, God had revealed to him that the Arya Samaj leader Pandit Lekh Ram, "shall be cut into pieces" like a calf. On March 6, 1897, Lekh Ram was stabbed to death by an assailant who fled the scene. Police searched Ahmad's property but found no incriminating evidence. This prompted a worried Dr. Henry Martyn Clark of CMS to have Ahmad charged with planning his own murder. It was alleged that Ahmad had deputized one of his disciples, Abdul Hameed, to murder Dr. Clark. Despite the intense sectarian atmosphere of Punjab, the case against Ahmad produced an unprecedented occasion of inter-religious harmony with orthodox Muslims, represented by Muhammad Husain Batalavi, and the Hindu Arya Samajis, represented by Chaudhary Ram Bhaj Dutt, the President of the Arya Samaj, joining hands with Dr. Clark to enter pleas to support Dr. Clark. Ram Bhaj Dutt, a lawyer, even offered his considerable legal expertise free of charge. Unfortunately for the multifaith coalition, and fortunately for Ahmad, the scrupulous District Magistrate, Capt. M.W. Douglas, found that the chief witness in the case had been schooled in his evidence by Christian missionaries who worked with Dr. Clark, and he acquitted Ahmad. Ahmad died of natural causes on 26 May 1908, leaving behind a community of thousands who believed him to be a divinely inspired reformer. Today, millions of Ahmadis present on every continent of the globe continue to believe him to be so[29].

7.2 The Social Identity Change Entrepreneurship (SIE) Model and Ahmad's Religious Innovations

Around the time Ahmad grew up, many thinking Muslims viewed the Islamic community to be under threat from Christian missionaries who were "insinuating themselves into every nook and corner of the Muslim world, and with heaps of abusive literature distributed freely among the Muslims, Christianity was challenging the very existence of Islam" (Ali, 1937: 16). Similar to almost all other North Indian Muslim boys growing in similar environment, he came to internalize this view of the world. With his elder brother having already taken up the role of the good landlord, Ahmad decided to adopt the role of a spiritual healer open to the scion of a feudal family. When one reads Ahmad's books and numerous volumes of his oral sayings recorded by his followers, one is overwhelmed with the strength of his identity as a Muslim to the exclusion of all other identities. While the words Muslims, Mussalmans, Islam, ummah, and believers are frequently repeated on every page, one is struck by the complete absence of terms such as Indians, Punjabis, Northwest Indians, or any other regional or ethnic identity terms in reference to himself. Clearly, Ahmad strongly identified as a Muslim above and beyond any other identity.

As an ardent Muslim, Ahmad was extremely motivated to resolve the social problems facing the Muslim community. Being a son of a feudal family with ethnic connections with the Mughal dynasty, formal religious training, and accolades for his early articles championing the defense of Islam, he certainly had the credibility needed to withstand the blow back that comes as a result of leading social change that he was convinced was needed in order to fix the broken superiority myths of Muslims. There is little doubt that Ahmad's followers respected him for his creative genius in resolving problems of interest to the Muslim community in a way that would enhance the status of Muslims,especially in comparison to the out-group, namely, Christians. Here's how Ahmad's follower Muhammad Ali summarizes Ahmad's major contributions:

> Deep religious mysteries which had baffled human minds for centuries had been unraveled. The second advent of Christ, the tribulation of the Anti-Christ, the prevalence of Gog and Magog, the coming of the Mahdi and similar other topics were mysteries which affected the two great religions of the world, Christianity and Islam, both contending for the mastery of the world, and an inspired man was indeed needed to lift the veil from the face of these mysteries. Such a man was Mirza Ghulam Ahmad. He was gifted not only with inspiration to elucidate the deepest mysteries, but also with the faith and energy which enabled him to give a new direction to the dissemination of Islam, which had hitherto found the West deaf to its message. Christianity was out to conquer the Muslim world; in temporal matters it had ousted Islam, but in the spiritual domain Mirza Ghulam Ahmad made a bold start and gave a challenge to Christianity in its very home. (Ali, 1937: 65)

How Ahmad came to believe that God had told him of Jesus' death and burial in Kashmir illustrates the process through which many religious social identity entrepreneurs come to believe that their their social identity change ideas are divinely inspired. Ahmad's early articles and books show that Ahmad held traditional Ahl-e-Hadith beliefs regarding Jesus' ascension to the heavens. In fact, he defended those beliefs against the rationalist Sir Syed Ahmad Khan who had argued in his commentary on the Quran that Jesus escaped death on the cross and died of natural causes later (Khan, 1880). However, some time during the early 1880s, Ahmad came to believe that traditional Muslims beliefs were indefensible in arguments with Christians. For years, he thought about possible solutions to this conundrum all the while praying to God to provide him with guidance.

Ahmad describes the intense joy that he received when he finally understood "the true meaning" of the Quranic verses and Ahadith concerning Jesus. He talks about that joy as, "more delicious than a king gets from his throne." . Why didn't Ahmad believe that it was he himself who had discovered these ideas? After all atheists also have fresh thoughts. Ahmad's answer would probably be that being the source of all knowledge it is indeed God who gives the knowledge to the atheists as well. It's just that an atheist is too ungrateful to acknowledge God as the source of his insight whereas a true believer is too humble to claim that he has discovered something on his own. He for, instance, that when a doctor thinks up a way of healing that is beneficial to us the doctor's search for an answer is akin to an answered prayer that had asked the Higher power for His blessing. "Even the people who have no connection to Allah or a belief in his existence also seek the unknown... but they do not know the source of their knowledge".

Ahmad had reasons to believe that God was talking to him, as he so desperately wanted God to talk to him, to help him save Punjabi Muslims from conversion at the hands of Christian missionaries. He preferred to live in a world in which he believed that God talked to him rather than living in a world in which he did not believe that God talked to him. Ahmad frequently asks such rhetorical question as "why was I chosen by Allah to have these unique insights? Why has he continued to shower his blessings on me if I am a liar?" The only answer to these questions according to Ahmad was that it was indeed God who had chosen him and given him such brilliant insights.

7.3 Summary

Mirza Ghulam Ahmad, the founder of Ahmadiyya Muslim Jama'at, was born in the Punjabi village of Qadian about 1839. He grew up as the youngest son of a family whose fortunes had waned along with those of the Mughal empire centered in Delhi. While his father and older brother busied themselves with managing their family's feudal lands, Ahmad developed his identity as an inheritor of the family's spiritual traditions of piety and healing. He became increasingly concerned with existential threats of conversion of Punjabi Muslims by both Hindu revivalist movements and

Christian missionaries. Of all the Muslim reform movements attempting to respond to these threats, his was most closely associated with the Ahl-e-Hadith movement that distinguished itself by (a) its vehement opposition to all sources of Islamic tradition except Quran and Hadith, and (b) its revival of the ijtehad doctrine which decreed that knowledgeable individual Muslims could themselves figure out solution to issues not directly addressed in Quran or Hadith rather than blindingly following traditions. This freedom not only allowed Ahmad to develop his innovative solution to various problems facing Indian Muslims but also allowed his fellow Ahl-e-Hadith to make sense of his claims that contradicted more than a thousand years of Islamic consensus. Ahmad argued that the only way to blunt Christian missionary arguments about the superiority of Jesus over Muhammad would be to change the traditional Islamic beliefs in physical resurrection of Jesus into a belief in a spiritual resurrection that occurred following his natural death. Appealing to Muslim hierarchy-enhancing myths, he argued that Muhammad was not only superior to Jesus but was a prophet-making super-prophet. Thus, Ahmad said that, unlike other people of the Book (e.g., Jews and Christians who also followed divinely-inspired prophets), Muslims could achieve all spiritual ranks open to man (including that of a prophet) if they followed Muhammad's path (i.e., Sunnah) - presenting himself as an example. He claimed that because of his close pursuit of Sunnah, God had made him a prophet who could show miracles to prove Islam's superiority over other religions. Since, Jesus had died and therefore could not have descended from heaven, prophecies of his second coming were to be fulfilled metaphorically through a Muslim who would be raised to the rank of a prophet. This Muslim was to be no other than the Mahdi. Ahmad claimed that God had told him that he was that promised Mahdi and Messiah.

Since to most non-Ahl-e-Hadith Sunni Muslims (as well as Shia Muslims and non-Muslims), Ahmad's claims appeared maximally socially counterintuitive, they immediately rejected him as a heretic and a false prophet. This is because they revered traditional authorities (including the four traditional Imams) and since Ahmad's claims clearly contradicted the tradition and were thus unacceptable to them. To some among the Ahl-e-Hadith, however, Ahmad's claims were minimally counterintuitive. This is because the Ahl-e-Hadith did not believe in the authority of the traditional imams, and they believed that Islamic scholars should be able to create new solutions to new problems. There is little doubt that many prominent Ahl-e-Hadith leaders (such as the founder of the Ahl-e-Hadith, Syed Nazir Hussain and his prominent student Muhammad Hussain Batalavi who coined the name of the movement) and their followers considered Ahmad to be an eminent scholar of Islam and they enthusiastically supported him – at least initially. Thus most of his early followers came from the Ahl-e-Hadith. The Ahl-e-Hadith leaders were, however, perturbed by Ahmad's stubborn refusal to tone down his claims to prophethood despite their repeated requests. They finally came around to denouncing him as a heretic decades after he made his initial claims, by which time Ahmad counted thousands of Ahl-e-Hadith lay people as his followers.

8 Ratcheting Up of Counterintuitiveness in Ahmadiyya Doctrine

8.1 Gradual Growth of Ahmad's Religious Innovations

Allah did not reveal all the mysteries to Ahmad at once. In fact, the process was far too slow and gradual, taking well over two decades and it only stopped with Ahmad's death (but as we'll see in the next section, Ahmadiyya religious innovation did not completely stop with Ahmad's death). This is not unlike the story of other religious innovators. But why should that be? Why don't religious innovators make all their claims once and be done with it? Why is the innovation process so deliberate and systematic with each major innovation followed by years of gestation during which time the innovation is discussed and elaborated (with decreasing amount of enthusiasm). Just as the extreme excitement and frenzy of the old innovation wears off and it becomes routine, a new innovation arrives to take its place. In order to answer such questions, we'll need to revisit the individual memory and reasoning processes that take place when people encounter novel and counterintuitive information outlined in Chapter 4 and how individual processes interact with group processes of ideology development outlined in Chapter 5.

Ahmad's first claim, namely, that of Jesus' natural death, was certainly culturally counterintuitive to most Punjabi Muslims, his primary in-group and target audience. Despite the fact that the idea had been floated by the highly-regarded Sir Syed Ahmad Khan in a footnote in his commentary on the Quran, it had never before been articulated with the kind of rationale in Muslim social identity that Ahmad had created (Upal, 2015c). The reaction by non-Ahl-e-Hadith Muslims ulema was immediate. They accused Ahmad of heresy and blasphemy for insulting Prophet Jesus and declared Ahmad an infidel.

As the context-based model developed in Chapter 4 indicates, when people hear or read an idea that runs counter to their ideology, they search their memory for knowledge that would allow them to justify the expectation violation. If they have acquired enough knowledge from their experiences to allow them to explain the cultural counterintuitiveness then they remember the new idea well. In addition, if the idea can be justified as being a consequence of their group identity beliefs, then they may be able to internalize the idea and commit themselves to it. Some Ahl-e-Hadith Muslims, such as Moulavi Nur-ud-Din, who were able to justify Ahmad's claim about Jesus' death on the grounds that that had been the true teaching of Islam all along and that belief in Jesus being alive in heaven had been a distortion of the true teachings of Islam. They were able to justify and make their Muslim identity coherent with the claim that Jesus was dead. The new identity of "true Muslim" that they had created as a result of this elaboration process was perfectly coherent with a belief in a dead Jesus (indeed "it demanded it", as Ahmad put it). Since - as a part of the justification construction processes the counterintuitive claim and positive aspects of social identity are activated and accessed in close succession - a close association is created between the counterintuitive claim and the group identity especially in the minds of group members but also to a lesser extent in the minds of other peoples. Thus, any mention of Jesus reminds Ahmadiyyas of their identity as members of the Ahmadiyya community and to a lesser extent non-Ahmadi Muslims who have had to justify their beliefs in encounters with Ahmadis may also be forced to think of Ahmadis when Jesus' death is mentioned.

When Ahmad ratcheted-up the social counterintuitiveness by making his next surprising claim of being a Messiah, it was seen differently by those from inside the community and those outside the community. Those inside the community who had already mentally justified his old claim of Jesus being dead found the new claim easily justifiable. If the old Jesus had died then the Messiah that the prophesied prophet Muhammad will appear at end of the 13th Islamic century, must be a new person. Ahmad's arguments that his becoming a prophet like Jesus by following Muhammad's teachings actually enhanced the status of prophet Muhammad over that of Jesus. In this way, denying Ahmad's prophethood would mean disrespecting prophet Muhammad. In Ahmad's (and his follower's) minds, both socially counterintuitive claims were logically implied by their belief that prophet Muhammad was superior to Jesus and which formed the core of their new "true Muslim" identity. This meant that both socially counterintuitive beliefs could be justified and made coherent by the community members. Beyond the individual memory processes, the social psychological processes outlined in Chapter 2 also support the incorporation of counterintuitive ideas into group identity. First, since the counterintuitive claims are supported by authority figures such as the group founders and/or leaders, the next generation learns that believing in these claims is an important part of group membership. Second, if the group members mostly interact with other in-group members, the positive relationship between counterintuitive claims and group identity keeps getting reinforced and over time the once counterintuitive beliefs no longer seem that way. Thus, believers in supernatural entities such as djinns, leprechaun, and ghosts do not think of these entities as counterintuitive. Instead, they hear about them and read about them all the time and may even start seeing them themselves. In short, such entities no longer violate their expectations. Once this happens, the old counterintuitive claims lose their excitement and the group becomes ready to receive new revelations.

In 1900, Ahmad announced his next major socially counterintuitive revelation that God had told him that "Jihad with sword has come to an end." This was a cancellation of a major pillar of Islam at a time when it was especially needed, namely, the period in which Indian Muslims were under British occupation. This was surprising to much of Ahmad's audience because up to this point he had championed the idea that Islam, unlike Christianity, was a religion to endure until the end of times, that not one iota of Quranic verses could ever be canceled. In fact, this had been his major line of attack against rationalists such as Sir Syed Ahmad Khan. Again, Ahmad appealed to the community's core mission of showing Islam's superiority over Christianity as a justification for his socially counterintuitive claim. He suggested that Christian missionaries had been maligning Islamic concept of jihad as being synonymous with fighting and by suggesting that Islam had been spread purely by the use of the sword.

In order to counter the Christian propaganda, Ahmad argued that he needed the kind of financial support that Europeans and Americans were providing to Christian missionaries. The early source of funding for his books had been feudal Muslim nawabs from throughout India. After establishment of the Ahmadiyya community in 1889, he asked his followers to donate 1/16th of their income for the propagation of his message. However, he felt that this was not enough and that more funds were needed if the message of a reformed Islam were to be propagated not only to Muslims in India and the Middle East but also to Christians in Europe and America. He proposed a solution in his 1905 book titled "al-*Wasiyyat*" (meaning the Will). The scheme essentially offered Ahmadis a burial ground in Ahmad's family graveyard (which he renamed *Bahishti Maqbara*, meaning a heavenly cemetery) in Qadian in exchange for a tithe involving donation of /10th of the value of all their properties to a newly-established association. *Sadr Anjuman Ahmadiyya* was to spend this money in maintenance of the cemetery and missionary work in India as well as overseas. The idea of selling people a piece of heaven seemed counterintuitive to most Indian Muslims and they vocally ridiculed Ahmad for proposing it (An-Nadawi, 1974). Ahmad fully anticipated such reaction. He wrote:

> It is possible that some people, who have an overwhelming tendency to think ill of others, may make me a target of criticism on account of this plan, believing this scheme to arise out of selfish motives, or calling it an innovation (bid'ah)...
>
> His [i.e., God's] purpose to separate the hypocrite from the believer by means of this plan... This trial is absolutely insignificant. The Companions of the Holy Prophet Muhammad were tried by being required to sacrifice their lives, and they gave their heads in the way of God...
>
> In the time of the Holy Prophet Muhammad, God the Most High had instituted some minor tests as well. For instance, it was the practice that no one could have a consultation with the Holy Prophet without first tendering a small offering. Thus, there was a test for the hypocrites in this as well. I myself feel that, by means of the test of this time too, those who are sincere to a high degree, who have in reality given priority to the faith over worldly interests, shall be distinguished from other people. This plan shall undoubtedly weigh very heavily upon the hypocrites, and it shall cause them to be exposed. And after death, they shall never be buried in this graveyard, be they men or women: "In their hearts is a disease, so Allah increased their disease" [Holy Quran, 2:10]. (Ahmad, 1905a: 32-33)

Those on the outside of the community who had not been able to justify Ahmad's old claims would have found that the addition of this new claim made Ahmad's ideology even more puzzling and even harder to believe. This is indeed what seems to have happened. The number of his opponents and the intensity of their opposition increased as Ahmad made more claims.

8.2 Evolution of Ahmadi Doctrine After Ahmad & The Big Split

Broadly speaking, Ahmad attracted two different kinds of people to his mission: well-educated urban Muslims who were primarily attracted by his championing the sectarian cause of defending Islam against Christians and Hindus through seemingly rational arguments, and devoted rural followers who saw him in the mold of charismatic central Asian saints who, according to Indian Muslims traditions, had converted thousands of Hindus to Islam through their supernatural and miraculous powers. The urban elite trained in the British-run educational institutions wanted him to be a reformer with the divine authority so that he could rid Islam of hard-to-defend and irrational practices of "Jihad with a sword" and the idea of "Jesus having been lifted up heaven and about to physically descend from heaven to earth." The rural devotees wanted a charismatic divine with miraculous powers (superior to the village sufi) so that they could interact with a living God and at the same time feel like they were part of a larger Muslim community outside their village. As long as Ahmad was alive, he was able to provide just enough to each group. The rational urban Ahmadis were extremely fond of his abrogation of Jihad with the sword, his killing of Jesus, and his frequent reformist exhortation against ignorant mullahs who gave sick people amulets and tawiz papers. The rural devotees loved his divine visions in which he often foresaw future events and his claims of having cured people through the power of his prayers. On the one hand, he claimed to be a divine prophet (superior to Jesus) but on the other hand, he claimed to be a reformist Muslim who was more devoted to the prophet of Islam than anyone else. However, after his death the two groups feuded with each other, forcing Nur-ud-Din to spend significant time and attention trying to keep them from calling each other non-Muslim.

8.2.1 Nur-ud-Din as a Successor to Ahmad

Nur-ud-Din had a unique stature in the community because of his status as having been the first in almost every desirable category. He was the first to accept Ahmadiyyat. He was Ahmad's foremost financier especially in the early years and his most trusted adviser and confidant. This meant that neither the rural nor the urban Ahmadi party could challenge his authority and his word was final.

The rural devotees with their traditional belief in the divine essence of a man being passed to his eldest son, looked with devotion to Ahmad's oldest son Mirza Basheer-ud-Din Mahmud Ahmad. Khawaja Kamal-ud-Din-the lawyer-and Moulavi Muhamad

Ali, MA, became the leaders of the urban educated community. All three men were members of Sadr Anjuman Ahmadiyya established by Ahmad before his death. Ahmad had also appointed Muhammad Ali as the founding editor of the *Review of Religions*, the primary Ahmadi mouthpiece to the English speaking world. He was also given the task of producing community's first translation of Quran into English. Khawaja Kamal-ud-Din, besides being a successful lawyer, was such an extremely talented orator that he became Jama'at's preacher-in-chief visiting fledgling Ahmadi communities all over India and holding meetings attended by thousands. In fact, it was the wild success of these meetings that so bothered Ahmad's oldest son, Basheer-ud-Din, that he set upon a tour of various Indian cities himself. However, the trip wasn't nearly as successful as that of Khawaja Sahib and Basheer-ud-Din argued that this was due to the less-than-honourable preaching strategy employed Khawaja Sahib.

> Khawaja Sahib would studiously avoid all mention of the Promised Messiah as, even when reference to him was called for by the subject of his address. He fully realized that without following such a course he could acquire no popularity among non-Ahmadis. And, as the objection of non-Ahmadis was to the personal claims of the Holy Founder of the Ahmadiyya Movement, the policy adopted by Khawaja Sahib brought him large audiences. Often these swelled into thousands, all eager listeners to what he had to say. Also, as has already been pointed out, Khawaja Sahib took special steps in order to make his lectures popular. The result was that his lectures achieved immense popularity and began to be applauded highly even by non-Ahmadis. Invitations for his lectures, began to pour in from all sides. When Ahmadis saw this eagerness and interest on the part of non-Ahmadis, many of them failed to realize its proper significance. What was only Khawaja Sahib's personal popularity began to be mistaken by them for the popularity of the Movement. Ahmadiyya Associations at various centres—either of their own accord or at the instance of Khawaja Sahib—began to arrange for special lectures by him. They did so under the impression that through these lectures non-Ahmadis would be brought nearer to the Movement, and would ultimately enter its fold. The epidemic became so prevalent that other lecturers of the Movement also began to adopt the same policy. It seemed imminent that the trumpet which God had sounded through His Messenger was to cease to resound forever. The time was one of extreme danger for the Movement. Some Ahmadi lecturers began to feel reluctant about making any reference to the Promised Messiah[as] in their lectures, and even when questioned on the subject, they tried to put it off by vague replies. This was not due to fear or hypocrisy, but the lecturers, following the example of Khawaja Sahib, had come to the honest conclusion that the adoption of his plan would prepare the way for the propagation of the Movement...
>
> I observed that everywhere members of the Jama'at were anxious that the lectures should contain no reference to Ahmadiyyat. For, they were afraid that in that case people would either not attend the lectures or else would receive them in a hostile spirit...One man from Lucknow who has since entered into my Bai'at actually wrote to Hadrat Khalifatul Masih I (ra) 'Would that you had sent along with the deputation a man of ripe experience, for these men adopt a method of preaching from which troubles may well be apprehended. Miyań Sahib (i.e. the present writer) is still in his early youth and is apt in his enthusiasm to overlook the suitability of the hour. God forbid that there should occur some violence and our name be brought to odium. There is still time for you to send a person of experience well aware of the need of the times'...

> At Lucknow I was originally proposed to deliver two lectures. But after one had been delivered, the other had to be abandoned; and this was mainly due to objections raised by Ahmadis themselves... The same was our experience in Benares... In short, people of the Jama'at in those days were undergoing a peculiar probation. On the one hand, in their heart of hearts they were aware that if they persisted in this policy of avoiding an open mention of the Promised Messiah[as],then the Movement would soon come to an end. On the other hand, after their experience of the practical success which attended the methods of Khawaja Kamaluddin, they felt afraid that the people would refuse to attend any lectures bearing on the distinctive claims of the Movement. (Ahmad, 1924: 259-280)

Thus Basheer-ud-Din argued that urban Ahmadis led by Khawaja Kamal-ud-Din were intentionally toning down Ahmad's religious innovations to gain acceptance among mainstream Muslims. This strategy, according to Basheer-ud-Din, ran the catastrophic risk of losing the movement's distinctive identity. Moulavi Muhammad Ali, on the other hand, argued that it was the group led by Basheer-ud-Din that had ratcheted-up religious innovations and was engaged in religiously prohibited innovations (*bid'ah*) in pursuance of their "sectarian narrow mindedness" (Ali, 1918). After talking about his unsuccessful tour of various Indian cities following Khawaja Sahib's successful tour, Basheer-ud-Din analyzes the differences among the communities in Lucknow and Cawnpore. He says:

> The reason why so few attended my lectures at Lucknow and Benares and so many attended them at Cawnpore seemed to lie in the fact that at the first two places I was unknown to the people whereas at Cawnpore, on account of the presence of many Punjabi traders, a section of the people knew and respected our family. This induced them to come and attend the lecture, and then truth was able to enter their hearts and their former regard was further enhanced. (Ahmad, 1924: 282)

Ahmad had made his aristocratic background as the scion of an ethnically Persian family as one of the factors that supported his claim to Messiah-ship since, in his interpretation, Prophet Muhammad had prophesized that the Messiah would come from among the Persians. When Ahmad found out that Government of India's 1901 census report in the Civil and Military Gazette had mistakenly identified Ahmad as the leader of a group of "street-sweepers" who mostly hailed from untouchable Hindus, Ahmad was incensed. He directed his associates to immediately send a protest letter to the government indicating that, "Government should know that sweepers are a criminal lot and we have never had any connection with them... This is a severe attack on my honor... Not only sweepers, I don't even have anything to do with those who are low class Muslims and have baser instincts" (M. G. Ahmad, 1903). Ahmad also claimed exception from rules for his family. In his 1905 book laying out the conditions for the *wasiyyat* system, he claimed that he and his family had been granted an exception from having to pay a tithe and follow other restrictions in order to be buried in the heavenly cemetery.

> God has made an exception in my case and the case of my wife and children. All other men and women must comply with these conditions; and whoever objects will be a hypocrite. (Ahmad, 1905a: 40)

Accusation of hypocrisy (*infa'q*) is one of the worst forms of insults for a Muslim given the rough treatment hypocrites (*munafiq'un*) receive in the Quran. No wonder, then that most of the rural devotees, who dominated the Jama'at in Qadian (and Cawnpore given Basheer-ud-Din's description), did not want to be counted among the hypocrites. Their interaction with Ahmad's family in general, and with his oldest son, Basheer-ud-Din, in particular, was no different than their interactions with traditional feudal lords and *sufi pirs* i.e., with great reverence. Abbaji, my maternal grandfather, definitely didn't want to be counted among the hypocrites. I remember, how he would fly into a fit of anger at any perceived insult of the hundreds-large *khanidan* (i.e., family) that inhabited a special central walled area of Rabwah. Our non-Ahmadi school teachers in Rabwah used to refer to Ahmad's family as "Rabwah's Royal Family" and they weren't far off from the truth since all of Ahmad's successors (i.e., Khalifas) except Nur-ud-Din have come from Ahmad's sons, grandsons, and great grandsons. I also remember how hard it was to get an audience with Hazoor. One would have to write repeated requests for months in advance to get an appointment to see him for a minute or two. Once the day came, you'd have to stand in line for hours outside the *qasr-e-khilafat* (the Palace of Caliphate) to get a chance to kiss Hazoor's hands (who at the time was Ahmad's eldest grandson) and how privileged you felt after having done so.

Among my mother's most prized possessions, next to the postcard bearing my name written by the third Khalifa (discussed in Chapter 1) lies the *sherwani* (a traditional Indian men's jacket) that the 3rd Khalifa gave to my father as a gift. This happened in 1981 when Abbijan was a missionary in Islamabad posted at Aabpara mosque/mission-house[30]. Hazoor used to stay at his villa (called *Bait-ul-Fazl*) in Islamabad's posh F-7 district, whenever he visited Islamabad (usually to apply for a visa for a foreign visit). During Hazoor's stay in Islamabad, instead of leading the evening prayers at the Aabpara mosque, my dad and I would hop on the Sohrab bicycle, issued to him by the Jama'at, to offer our ritual evening prayers with Hazoor. After the evening prayers, I'd bike home by myself while my father would stay behind along with dozens of other devoted Ahmadis waiting to catch a glimpse of Hazoor going from one meeting to another. If they were really lucky, Hazoor would send them his leftover food and drinks. Since, this *tabarruk* (literally source of blessings or barakah, but refers to the food or anything else that touches the hands or lips of the Khalifa) was considered highly desirable, he would be very happy on the days when he got a piece of it. I vividly remember – as if it was yesterday - the day he got the sherwani from Hazoor because it is by far my most tender memory of him. To give you an indication of how much the security situation in Pakistan has deteriorated in the last 30 years, I, a boy of 11 at the time would bike from one end of Islamabad to another in the dark of night without any fear. On that day, when I got back to Aabpara, I found my friends Kiki and Thai-Thai, two Chinese boys,whose parents were chefs at the Ambassador Hotel next door, playing soccer in the long narrow veranda outside the hotel. I joined them in the game. One of us hit a shot that landed under the huge metallic natural-gas meter at the edge of the veranda. As I picked up the ball, I hit my head against the meter really hard. It hurt badly. I thought that I had gotten a cut because it stung and felt wet when I touched it. I stopped playing and went to bed to sleep, deeply scared that Abbijan would find out that I was playing rather than doing my homework. The fear returned when I woke up in the middle of the night with Abbijan sitting by my pillow and combing his fingers through my hair. Why was he doing that? I wondered. He had never done that before. 'Oh, no', I thought. He knows. I am in trouble! I felt a palpable sense of relief when he told me that this was the happiest day of his life because Hazoor had given him his very own sherwani. Since my father was significantly slimmer than Hazoor, refitting the jacket left him with pieces of tabarruk cloth that he distributed to our immensely grateful relatives.

If this is how Ahmadis treated Ahmad's grandson, I can only imagine how much they must have revered Ahmad's eldest son. Being a smart man, Basheer-ud-Din must have known that his status was derivative of Ahmad's position as creator of the distinctive Ahmadi doctrine and founder of the distinctive Ahmadiyya community. Thus Basheer-ud-Din clearly had an interest in emphasizing Ahmad's innovations, his unique status, and other distinctive features of the Ahmadiyya beliefs and organization.

Khawaja Kamal-ud-Din and Moulavi Muhammad Ali, on the other hand, had no relationship with Ahmad other than the shared belief that Islam was in danger and needed to be reformed to be saved and revived. Even though, they had left their families and friends in Lahore to join the community in Qadian, they had to travel frequently back to Lahore and other cities dominated by non-Ahmadis. Their educational and worldly accomplishments in the British-Indian system were appreciated and respected by the rural devotees in Qadian but their interactions with the devotees were no substitute for the refined intellectual discussions that they had with their equally accomplished non-Ahmadi Muslim peers whom they met in these cities. They not only wanted to maintain those relationships but wanted to expand them so that they could play a role in the larger Islamic revivalist movements. They may even have had hopes of being considered for leadership by members of Indian Muslim movements such as the Anjuman Himayat-e-Islam or the All India Muslim League. They certainly had no interest in declaring their friends to be infidels and ceasing all social interactions with them.

One of the most significant events in the battle between the two camps happened in 1911 when Basheer-ud-Din published an article titled, "Why no one can be a Muslim without accepting all of God's prophets" in which he argued those Muslims who had not formally offered allegiance (bayah) to Ahmad were non-Muslims. Khawaja Kamal-ud-Din pleaded with Nur-ud-Din to rebut Basheer-ud-Din arguing that not doing so would endanger the lives of Ahmadis living in areas dominated by non-Ahmadi Muslims, especially those living in Pashtun-dominated areas of India's borders with

Afghanistan. Nur-ud-Din reluctantly signed on to a clarification which toned down Basheer-ud-Din's article by suggesting that non-allegiance pledging Muslims were only "infidels of Ahmadiyyat" and not "infidels of Islam."[31]

In 1912, both groups asked Nur-ud-Din for permission to start publishing their own newspapers. Basheer-ud-Din founded *Al-Fazl* ("the blessing") in Qadian and Moulavi Muhammad Ali founded *Paigham-e-Sulah* ("message of peace/compromise") in Lahore. Nur-ud-Din wanting to keep everyone happy gave his blessings to both groups. The publications gave both parties a chance to communicate their competing messages to the larger Ahmadi community and beyond to convince people of the truth of their cause. While *Paigham-e-Sulah* published glowing accounts of the success that Khawaja Sahib was having in India and England (where he moved in 1913 to open the first Ahmadi mission abroad) in attracting people to the message of Islam, *Al-Fazl* focused on the importance of the institution of Khilafat (i.e., Caliphate) and on doctrinal differences between Ahmadis and non-Ahmadis. Both sides wanted to win the majority of Ahmadis to their side when the time came to select a successor to the sick and elderly Nur-ud-Din. That moment came on Friday the 13 March 1914.

8.2.2 Election for Nur-ud-Din's Successor

The rural group had hoped that Basheer-ud-Din would be elected as the next Khalifa by a majority of Ahmadis while the urban group hoped that a majority would see that there was no need for the continuation of the office of Khalifa since the Anjuman was able to handle the business of spreading Ahmad's message of Islamic revival around the world. However, to the great disappointment of the urban group, Basheer-ud-Din was quickly accepted as the second Khalifa with only a few dozen Ahmadis refusing to offer him their oath of allegiance. This forced the urban-group to move to Lahore and establish a new association called the *Ahmadiyya Anjuman Ishat-e-Islam Lahore*. I will refer to those who remained in the village of Qadian as the Qadiani Ahmadis and to those who moved to Lahore as the Lahori Ahmadis

Two related points of debate and disagreement emerged among the Lahori and the Qadiani group: the prophethood of Ahmad, and necessity of beleiving in Ahmad as a precondition for one's salvation. On both those points Qadiani Khalifa Bashir-ud-Din ratcheted up counterintuitiveness. He declared unequivocally that Ahmad had been a prophet whose acceptance is required for salvation and that all non-Ahmadi Muslims are not Muslims and will thus will not be saved. These were not just dry theological points; they had deep implications for how Ahmadis were to interact with other Muslims. Bashir-ud-Din declared that Qadianis could no longer pray with non-Qadiani Muslims, marry them, or take part in their marriage or death rituals. This led to the famous incident in which Pakistan's first foreign minister Sir Muhammad Zafrullah Khan, who was a Qadiani Ahmadi, silently watched the funeral of Pakistan's founder, a non-Ahmadi Muslim, without participating in the funeralprayer led by a non-Ahmadi Muslim. On all these issues, the Qadiani position represented a clear departure from the positions taken by Ahmad himself who when questioned about his claim to prophecy in light of the orthodox Muslim belief in finality of Muhammad's prophethood, denied that he had ever claimed to be a 'full prophet.' Ahmadis had also been free to marry non-Ahmadi Muslims and take part in their social activities throughout Ahmad's lifetime.

By staying in Qadian, Qadianis had access to the Bahishti Maqbara (the heavenly cemetery) for burying their *moosies* (the people who subscribe to the wasiyyat system established by Ahmad in his 1905 book). Worried about falling *wasiyyat* revenues, however, in 1925, the Qadiani Khalifa Basheer-ud-Din argued that some changes were sorely needed. These included denying *wasiyyat* to those Ahmadis who didn't own enough property to contribute money and adding a monthly tithe as 1/10th of the monthly income over and above the 1/10th of the property that had already been instituted by Ahmad when he announced the scheme in 1905 (B. M. Ahmad, 1925). In 1947, when British India was partitioned into India and Pakistan and Qadian fell on the Indian side of the newly-drawn international border, Qadiani Ahmadis decided to move to Pakistan leaving behind 313 men (the number of Muhammad's companions in the epic battle of Badr) to protect their holy sites. Initially, the Qadiani Ahmadis led by their caliph Basheer-ud-Din moved to Lahore. For a few years, both Ahmadi factions were headquartered in the same city. Qadiani moosies who died during this time had to be buried in ordinary graveyards, similar to the fate of Lahori moosies. Both factions invoked Ahmad's 1905 extra-ordinary circumstances clause to justify such burials.

> If a person bequeaths one-tenth of his property and then he happens to meet death by drowning in a river, for instance, or he dies in a foreign land, from where it is difficult to bring back his body, his will shall still stand valid, and with God it shall be as though he were buried in this graveyard. (Ahmad, 1905a)

In other words one could enjoy all the benefits of the heavenly cemetery without being buried in one!

Qadiani Ahmadis searched all over Pakistan looking for suitable land to establish their movement's new headquarters. After looking at various parts of Pakistan, a barren chunk of land on the western hilly bank of Chenab River was chosen by Basheer-ud-Din because of a vision, not unlike the vision that had led Ahmad to establish Bahishti Maqbara in the first place. All of the Ahmadiyya institutions left behind in Qadian were recreated in Rabwah including the Sadr Anjuman, Masjid Mubarik, Masjid Aqsa, Talim-ul-Islam School, Jamia Ahmadiyya, and Bahishti Maqbara.

One wonders how many Ahmadis would have justifyied the claim that burial in a tract of barren hilly land thousands of miles west of Qadian in the middle of nowhere would guarantee one paradise in an afterlife even if Ahmad had made such a claim in 1905. That idea would certainly have been maximally socially counterintuitive to

Ahmadis, had Basheer-ud-Din (a boy of 16 then) had made it in 1905. Yet once the idea of the original Bahishti Maqbara had been accepted, Basheer-ud-Din's status as a divinely appointed Khalifa recognized, and the migration of the Ahmadiyya headquarter from Qadian to Rabwah had been justified as religiously sanctioned *hijra* (Arabic term usually used to refer to Prophet Muhammad's migration from Mecca to Medina), most Ahmadis were able to make sense of the claim that the lot of land in the hills of Rabwah was also a heavenly cemetery. Indeed, record shows that establishment of Bahishti Maqbara in Rabwah was a non-issue for most Qadiani Ahmadis. Similarly, the recent proliferation of Bahishti Maqbaras in the West (e.g., UK & Canada), following the mass migration of Ahmadis to the West in the last few decades, has hardly raised any eyebrows among Qadiani Ahmadis. On the other hand, Lahori Ahmadis see the heavenly cemeteries in Rabwah, Canada and UK as maximally counterintuitive.

The fears that Khawaja Kamal-ud-Din had expressed to Nur-ud-Din in 1911 were realized as Ahmadis came under attacks from mobs of non-Ahmadi Muslims in Pakistan in 1953 and then in 1974. However, the theological differences between the Qadianis and Lahoris did little to spare the lives of Lahori Ahmadis who lost a proportionate number of lives in the anti-Ahmadiyya riots of 1953 and 1974. To quell the 1974 riots, Prime Minister Z. A. Bhutto amended Pakistan's constitution to declare all Ahmadis non-Muslims without distinguishing between Lahoris and Qadianis. The 1984 Martial-law order (passed by parliament and added to the constitution in 1985) that made it illegal for Ahmadis to engage in any Muslim practice such as greeting people with an "*Assalam-o-allaikum*"(Peace be Upon You) or reciting *adhan* (the call to prayer) also explicitly included Lahoris. Khawaja Kamal-ud-Din's strategy of toning down Ahmad's claims did not succeed in appeasing non-Ahmadi Muslims. One may ask whether or not that would have happened if all Ahmadis had adopted Khawaja Kamal-ud-Din's position i.e., if Qadianis had never existed. Would Ahmadis have survived and prospered as a distinct community if all Ahmadis had intermarried with Sunni Muslims and not just the Lahori Ahmadis? Would the resulting stronger social ties have put breaks on anti-Ahmadiyya rhetoric in India and Pakistan? If the highly visible symbols of Qadiani Ahmadiyya distinctiveness (such as the city of Rabwah, its "heavenly cemetery," the image of Sir Zafrullah Khan sitting idly by while Sunni Muslims offered funeral prayer for Pakistan's founding father Muhammad Ali Jinnah) did not exist, would anti-Ahmadiyya forces hate Ahmadis any less? We will probably never know answers to all these fascinating questions and therefore will never be able to know which strategy would have been more beneficial to the Ahmadis in the long term.

At the heart of the birth of many new religious movements lies a novel idea that seems counterintuitive to most members of the larger parent community from which the new community draws most of its members. According to the context-based model, once this counterintuitive idea becomes embedded within the new group's ideology, it is no longer seen as counterintuitive by members of the new community. This results in the loss of memorability and transmission advantages for the new idea among members of the new community. The social identity entrepreneurial members of the new community (such as Kamal-ud-Din & Bashir-ud-Din) must either build upon the counterintuitiveness of the old idea or remove some of its counterintuitiveness in order to catch people's attention in the cultural environment of the new NRM. The adoption of such conflicting strategies can sometimes result in the breakup of the original movement (labeled P in Figure 9) into sub-movements (labeled C– & C+ in Figure 9).

An analysis of the split in the Ahmadiyya Jama'at allows us to see how the two strategies of ratcheting-up counterintuitiveness and toning it down played out in the real world. As Lahoris found out, toning down the counterintuitive ideas of the movement's founder also removed some of the group's distinctive aspects. This has the effect of making the group (C– in Figure 9) less distinct from the grandparent community (G in Figure 9). This not only runs the risk of the group merging back into the grandparent group but it also makes it less attractive to potential recruits from the grandparent community because the group's ideas are no longer counterintuitive to members of the grandparent group. This makes toning-down of counterintuitive ideas of the founding fathers a less attractive proposition for new religious movements and explains why many NRMs adopt the strategy adopted by Qadiani Ahmadis and double down on the counterintuitive ideas of their founders.

The strategy of ratcheting-up counterintuitiveness, however, is not completely risk-free either. While the new doubly counterintuitive ideas are minimally counterintuitive to the members of the NRM (e.g., the community labeled P in Figure 9), they could become so far-fetched from the shared beliefs of the grandparent community (labeled G in Figure 9) that they may seem maximally counterintuitive to most members of that community. This would mean that while the new community (labeled C+ in Figure 9) attracts members from its parent community (labeled P in Figure 9), it has increasing trouble attracting members from the much larger grandparent community (labeled G in Figure 9) as it drifts further and further away from the mainstream. The Qadiani strategy of emphasizing points of Ahmadiyya distinction from Sunni Muslims (i.e., Ahmad's prophethood and *takfir* of non-Ahmadi Muslims) risked making the Qadiani Ahmadi belief system too counterintuitive for Sunni Muslims to comprehend. A way to escape this bind would be for the community holding a doubly counterintuitive belief system to adopt two different missionary strategies: emphasizing doubly counterintuitive beliefs when interacting with fellow NRM members (i.e., members of community P/C– or fellow Ahmadis/Lahoris in case of Qadianis), and de-emphasizing doubly-counterintuitive beliefs when interacting with non NRM members (such as members of G or the Sunni Muslims in case of Qadianis).

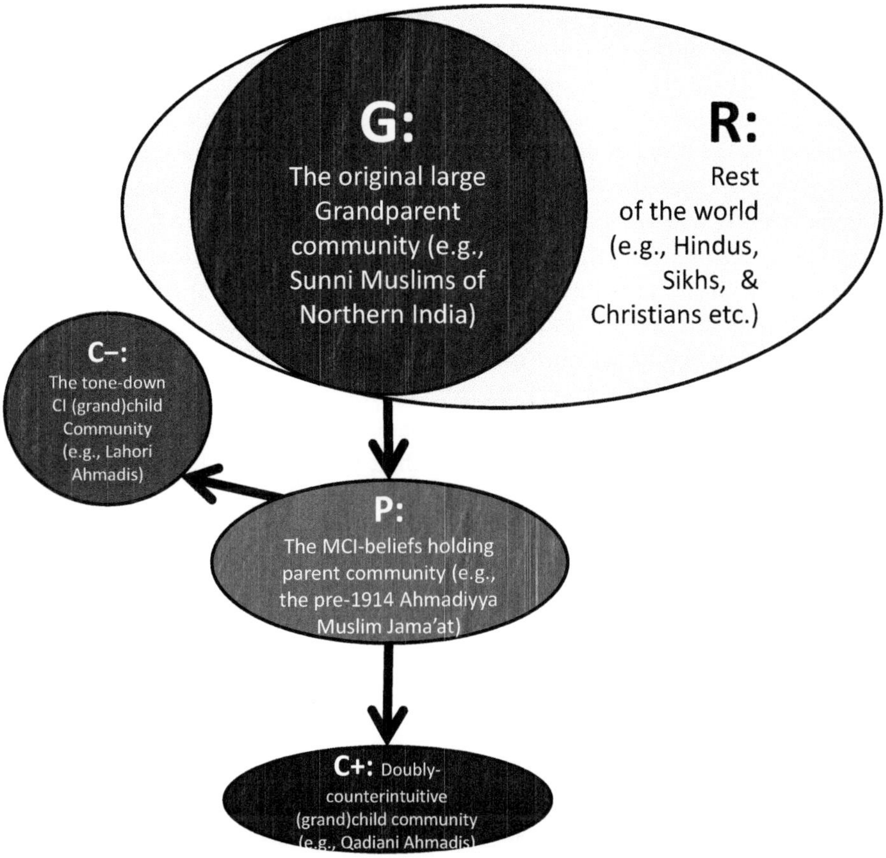

Fig. 9: A Visual representation of ideological splits in new religious movements. G is the original community from which an NRM (P) splits up because of the counterintuitive claims of a social identity entrepreneur. The movement P splits into two subgroups C– and C+: when social identity entrepreneurs (who are members of P) adopt two different strategies for attracting members to their faction. The leaders of C+ ratchet up counterintuitiveness while leaders of C– tone it down.

This analysis also suggests that counterintuitive beliefs should be even further de-emphasized when interacting with the rest of the world (i.e., R in Figure 9). Thus Qadiani Ahmadis should adopt a branding strategy similar to that advocated by Lahori Ahmadis when dealing with non-Ahmadis. This is indeed what we find when we look at the branding strategy adopted by Qadiani Ahmadis in the West (since the movement of their headquarters to the UK in 1984). We find that this strategy is virtually indistinguishable from the one successfully adopted by Khawaja Kamal-ud-Din at the start of the twentieth century. This was the strategy that caused much consternation among the rural Ahmadis at the start of the twentieth century and which eventually lead to the split in the movement. The basic tenet of the strategy is that Ahmadis should be seen as moderate, reasonable, and intellectual Muslims with whom one could discuss a disagreement over a cup of tea (as opposed to irrational fundamentalists or the tomb-worshipping traditionalist Sufi Muslims). As Basheer-ud-Din argued in his criticism, this strategy seems dishonest to the extent to which it downplays the doctrinal disagreements that Ahmadis have with mainstream Muslims on issues - other than Jihad. Thus while Ahmadis are keen to emphasize how they disagree with Muslims on the legality of Jihad against the West, they underemphasize their differences on the finality of Muhammad's prophethood. This is because while abrogation of Jihad places them on the same side as moderate Muslims around the world, belief in a prophet after Muhammad does not. Even though this strategy is wholly reliant upon the ignorance of Western media and political class about Islamic and Ahmadiyya doctrine, it seems to have worked well so far. Recent events such as the establishment of the Caliphate by Jihadists of the Islamic State has handed Ahmadis a golden opportunity to contrast their peaceful "Khalifa" (whom they've recently taken to calling the "Khalifa of Islam") with Islamic State's violent Khalifa Abu Bakar Al-Baghdadi. Thus Qadiani Ahmadis have become the favourite Muslims of politicians and media across the political spectrum in the West (especially in UK and Canada) as evidenced by recent statements by Western politicians:

> This [Ahmadiyya] community is a model of integration and presents a true authentic view of a peaceful Islam that is at home here in Canada" (Canada's Minister of Citizenship and Immigration, Jason Kenney at the Opening of Ahmadiyya Mosque in Vancouver, 18 May 2013).

> In this [Ahmadiyya] mosque, we see the love and benevolent faith of Islam. Ahmadis embrace Canada and Canada has embraced you. (Canada's Prime Minister Steven Harper at the Opening of Calgary Ahmadiyya Mosque on 5 July 2008; Calgary-Herald, 2008)

> [Ahmadiyya] philosophy of 'Love for All and Hatred for None' is exemplified by its condemnation of the horrific murder of Drummer Lee Rigby in Woolwich." (British Home Secretary Theresa May speaking in the House of Commons; HC, 11 June 2013)

> Britain's Ahmadiyya Muslims work hard and contribute greatly to this country. Their belief in peace and religious tolerance is an example to us all, and is to be expected from a community whose motto is, "Love for all and hatred for none... They reject terrorism in any form. Ahmad declared that jihad by the sword had no place in Islam. Instead, he wanted his followers to wage a bloodless, intellectual jihad of the pen to defend Islam... In Britain today, we regard such attributes as modern and tolerant. (British MP Siobhan McDonough speaking during a House of Commons Debate; HC, 20 October 2010)

Perhaps given the effusive praise by these leaders, it is not surprising that one of the most frequent questions I am asked when I discuss Ahmadiyya Muslim Jama'at in a Western audience is "aren't Ahmadis moderate Muslims?" My answer is both yes and no. While Ahmadis are clearly moderates on the issues of jihad and terrorism (perhaps a deciding issue for those, such as the US President Donald Trump, who perceive Islamic terrorism to be the biggest threat facing the West), the positions they hold on the status of women (e.g., the requirement to wear a veil covering a woman's face and a man's right to have a maximum of four wives), the rights of gays and lesbians (e.g., opposition to same sex marriage), and freedom of speech (e.g., punishment for blasphemous speech and the Khalifa's power to expel Ahmadis from the Jama'at for "disobedience") are diametrically opposed to those advocated by moderate liberal Muslims such as Maajid Nawaz (Nawaz, 2012), Mustafa Akyol (Akyol, 2011), and members of the Muslim Reform Movement (Richardson, 2015).

Mirza Ghulam Ahmad, for instance, claimed to be the first Muslim leader to initiate a petition (published on 22 September 1895) demanding that the British Indian government amend Indian Penal Code 298 to make it easier to prosecute anyone who blasphemes a founder of a major religion[32]. In 1927, Ahmadis took the lead in agitating against publication of the book *Rangila Rasul* by Hindu author Raj Pal and demanded that the book be banned. One of the movement's most prominent members, Sir Muhammad Zafrullah Khan, also spearheaded legal action against the book. Defense of the honour of the Holy Prophet Muhammad has become such an integral part of the Ahmadiyya identity that Ahmadis have been at the forefront of demanding blasphemy laws around the world. When he spoke to the European parliament members, the current leader of the Ahmadiyya Jama'at demanded the same of the Western governments, "Governments need to make policies that establish and protect mutual respect, through which hurting the sentiments of others or causing them any type of harm should be outlawed" (Ahmadiyya-Gazette, 2013). Moderate fundamentalism is a better way to summarize Ahmadiyya Muslim Jama'at's positions on issues of concern to most Westerners.

A Pew survey conducted between 2008 and 2012 in 39 Muslim countries around the world (Pew-Research-Center, 2013) found that a majority of Muslims hold attitudes that are hard to characterize as either fundamentalist or moderate. For instance while majorities had an unfavourable view of ISIS and opposed terrorism against civilians, they did not support human rights for all. In 20 out of 23 countries where the question was asked, majorities agreed that women must obey their husbands. Majority of Muslims in Middle East, North Africa, and Pakistan said that women do not have a right to divorce. Majorities in all six regions of the world (South Asia, Southeast Asia, Middle East & North Africa, Sub-Saharan Africa, Southern Eastern Europe, Central Asia) surveyed by Pew also declared homosexuality, sex outside marriage, and abortion to be morally wrong. Thus moderate fundamentalism may also be a better description of the views of a majority of Muslims around the globe and not just Ahmadi Muslims.

9 Conclusion

> ...ever since the Age of the Enlightenment, leading figures in philosophy, anthropology, and psychology have postulated that theological superstitions, symbolic liturgical rituals, and sacred practices are the product of the past that will be outgrown in the modern era. The death of religion was the conventional wisdom in the social sciences during most of the twentieth century; indeed it has been regarded as the master model of sociological inquiry, where secularization was ranked with bureaucratization, rationalization, and urbanization as the key historical revolutions transforming medieval agrarian societies into modern industrial nations. As C. Wright Mills summarized this process: "Once the world was filled with the sacred – in thought, practice, and institutional form. After the Reformation and the Renaissance, the forces of modernization swept across the globe and secularization, a corollary historical process, loosened the dominance of the sacred. In due course, the sacred shall disappear altogether except, possibly, in the private realm." (Norris & Inglehart, 2011: 3)

While the secularization hypothesis no longer holds sway as it once did among scholars of religion, many scientists still find exhuberant religious passions, especially as displayed by many new religious movement leaders and their followers, to be puzzling. No wonder they are so attracted to the psychopathology model of the origin of religion (reviewed in Chapter 1) which suggests that excessive religious devotion is a result of one mental illness or another. A similar view, propounded by so-called cult-hunters, claims that cult leaders are morally defective people who seek to control other humans for their pleasure as they are unable to overcome their lower urges. They find impressionable people suffering from various emotional and cognitive impairments and abuse them. Families and societies need to band together to save their vulnerable members from being entrapped by the evil cult leaders using every tool available. Traditionally, the only moderating voices against such hysteria have been those of sociologists of religion who have advocated the need for a scientific approach to understanding new religious movements. Unfortunately, they lack the tools needed to develop a fully scientific account of new religious movements. This is because a scientific explanation of new religious movements requires an understanding of individual cognitive tendencies that cause some individuals to create and spread novel religious ideas and cause others to put their faith in them. A traditional sociological approach that solely focuses on sociological variables is grossly incomplete. Similarly a traditional psychological approach that studies a brain in a vat and ignores various contextual factors will not suffice. Instead, a fully multidisciplinary approach that relates sociological variables to the individual cognitive variables is required. A number of social scientists, including cognition and culture researchers, have advocated such an approach (Sperber, 1996). Some cognition and culture researchers, calling themselves cognitive scientists of religion (CSR), have advocated using this multidisciplinary approach to the study of religion. One of the early successes of the CSR approach has been the development of the minimal counterintuitiveness hypothesis (reviewed in Chapter 4) which explains that most of the religious concepts are minimally counterintuitive.

Unfortunately, traditional cognitive scientists of religion argue that the minimal counterintuitiveness hypothesis does not apply to the spread of counterintuitive NRM doctrine or any other culturally-learned ideas. They argue that this is a good thing because this frees up the scholars of religion from having to deal with messy things such as sociocultural and historical contexts in which an idea gains popularity in a population. A universal cognitive theory will be able to explain the success of religious ideas in all cultures for all times they claim. The context-based model of the minimal counterintuitiveness effect presented in Chapter 4-5 argues otherwise. It suggests that concepts that violate culturally-learned ideas also enjoy transmission advantages due to memorability and that context must be taken into account in order to understand whether or not a concept is counterintuitive for a given population. This view not only redeems an historical and sociocultural study of religion as an essential complement for a cognitive science of religion, it also provides a systematic way of integrating the two in developing a truly scientific approach to the study of religion. As illustrated by the detailed historical study presented in Chapters 6-8, historical accounts are necessary for an understanding of the context in which an idea was presented, and to determine its memorability advantages in a given population. The transmission advantages due to memorability are not enough to explain how religious concepts become embedded as part of a group's social identity. We need to look at social psychological processes of social identity formation in order to understand how that happens. The model presented in Chapter 3 focuses on the role that group leaders play in creating novel changes to a group's shared beliefs and selling these changes to their fellow group members.

The context-based model of the minimal counterintuitiveness effect and model of leaders-as-social-identity-entrepreneurs forms the core of the cognitive science of new religious movements (CSNRMs) developed here. By underemphasizing the role that mental illness plays in the religious claims of new religious movement founders and by outlining how ordinary processes of cognition can cause mentally healthy people to engage in such behavior, the CSNRMs points out a number of points of commonality -between so-called "ordinary" people and religious innovators and followers. According to CSNRMs, primary motivations of NRM founders and followers are social-psychological. Far from setting out to start new religions, NRM founders are concerned about the survival and prosperity of an existing religious group. As staunch group members, their natural instinct is to deny any purported shortcoming in their group's shared beliefs and behaviors but once they become convinced that a change is necessary for their group's survival and growth, they are willing to spend considerable resources in convincing their fellow group members of the need for change. Thus Mirza Ghulam Ahmad, the founder of Ahmadiyya Jama'at, sincerely believed that unless Indian Muslims changed their beliefs about Jesus and his second coming, they faced a catastrophic attrition to Christians. His son Basheer-ud-Din believed that unless Ahmadis continued to emphasize distinctive features of Ahmadiyya doctrine and organization, they were in danger of losing their unique identity and becoming mixed up with the rest of world's unorganized and leaderless Muslims who were destined to lose to the stronger and better-organized Christian West

Similar to other marketers, the social identity entrepreneurs have to create a narrative of change that is effective for their target audiences. Effective messages of social identity change have to be memorable and persuasive in order to achieve the desired impact on the targeted group members. Surprise is one of the especially effective techniques that marketers[33] – including social identity entrepreneurs – use to get the attention of their target group members. The context-based model of human memory presented in Chapter 4 suggests that this is so because evolution has favoured those memory architectures that allow people to spend more cognitive resources to process surprising ideas because such ideas indicate a gap in a person's world knowledge and present a learning opportunity. However, in order to learn from such an opportunity a person must have the background knowledge needed to fill the gaps opened up by the surprising information. Through the justification process, the target of a well-designed surprising message creates new knowledge that bridges the newly-opened gap. By familiarizing themselves with their target audience's mental knowledge and by carefully designing surprising messages, good marketers can ensure that the new knowledge created by their targeted audience is what they intended it to be. Social identity entrepreneurs such as Ahmad also have to ensure that the information created by their targeted group members, as a result of their surprising message, is in accordance with the group's myths (e.g., the group's superiority myths). Ahmad utilized his familiarity with beliefs of North Indian Sunni Muslims to design surprising claims such as Jesus-is-dead and I-am-a-Jesus-like-prophet and provided sufficient context to allow his targeted audience members to create the knowledge that this claim was needed to support the Muslim superiority myth that Muhammad is better than Jesus. Overtime, however, this once-surprising-message became a commonly-accepted truth among the Ahmadiyya community. According to the context-based model, once this happens, ideas lose their attractiveness and social identity entrepreneurs must build on the former surprise factor in order to continue to attract their target audience's attention. This explains both splits in religious movements (e.g., the Ahmadi split into Qadiani and Lahori groups) and the cultural success of maximally counterintuitive concepts, such as the Christian-Judaic and Islamic concept of God.

While, up to this point, I have emphasized similarities between secular social identity entrepreneurs (such as political and civic leaders) and religious leaders, there are also important differences between the two types of social identity entrepreneurs. For instance, while secular leaders are keen to take credit for their social-change messages, religious leaders claim God as the sole source of their ideas. To understand why this happens we need to understand how religious leaders develop their understanding of their role. To study the question of how people recognize their role in life, sociologists Peter Collier and Peter Callero designed an interesting study in which they asked 40 sophomores at a Portland high school to participate in a recycling education program (Collier & Callero, 2005). For next six weeks, eight students listened to lectures and engaged in group discussions about recycling. They went to a local garbage station, conducted field research about the extent of recycling in their neighbourhood, produced videos to promote recycling among elementary school students in which they were asked to say, "My name is ____ and I recycle", and made presentations to elementary school students to encourage them to recycle. The high school students who participated in the study, as well as some non-participating sophomores, were asked to answer a questionnaire before and after the study in order to better understand cognitive processes involved in adoption of a new role. The sociologists found that there were three key processes involved:

> First, there must be recognition and understanding of what it means to be a recycler. Recall that a cultural object emerges from shared meaning. The role will not be recognized or accepted as legitimate if meaning is not shared...
>
> The second step in the process of behavioral commitment requires that the recycler role be used as a resource for defining the self. Which is to say, the actor can begin to proclaim, "I am a recycler" with relative assurance. It is at this point that the role comes to serve as a basis for identity. Third, a correspondence between the meaning of the role and the meaning of the self develops as a consequence of using the role to define the self. At this point, the actions and attitudes associated with "recycler" are now also associated with the self." (Collier & Collero, 2005)

This means that to adopt the role of a religious innovator, a divinely-anointed leader (such as an Islamic prophet) one must intimately understand what that role entails. There is little doubt that Ahmad was well-versed in the intricacies of the Islamic doctrine regarding the role of a prophet. While growing up, Ahmad's favourite books included Islamic books about prophets and saints and various miracles performed by them. He also intimately knew what it meant to receive divine revelations because his older sister regularly received divine revelations from God. There are also suggestions that his first cousin, and best friend, Mirza Imam-ud-Din had founded a new religious movement (Ahmad, 1935). Ahmad wrote extensively about what it means to be a prophet who received revelations from God. As predicted by the role theory model, Ahmad's Muslim detractors who did not share his concept of a prophet existing after Muhammad questioned Ahmad's legitimacy as a divine messenger.

Having a source of knowledge about what a prophet does is not the only thing that a prophetic role model provides to an apprentice. The high regard in which the role model is held by the friends and family of the apprentice make the role model highly desirable. Emulation of the actions, behaviours, and even the physical appearance of the prophet are considered so desirable that those factors usually becomean explicit part of religious doctrine. Emulating the *sunnah* (i.e., reported behavior) of Prophet Muhammad is considered to be the only way to be a true Muslim because Quran does not completely specify details of good Muslim behavior. For instance, while the Quran does repeatedly ask Muslims to offer multiple ritual prayers every day of the week but it does not specify precisely how to do it. This was only shown by the Prophet Muhmmad. Prophet Muhammad is also reported to have said that “one cannot attain true faith until one loves me more than one’s parents.” (Bukahri: Book of Belief)

If you happen to be lucky enough to be better at emulating the role model than most of your peers, then in a society where that’s a desirable thing to do, it is only natural for your friends and family to complement you on your being so prophet-like in your behavior. This not only reinforces such behavior but it also amplifies the sense of being exceptional that all of us are born with. Every time you have a dream which comes true, and your friends and family tell you that you are divinely blessed, you wonder about the reason that God chose you for this special blessing. When you hit upon a novel solution that seems so good that it can overcome the monstrous problems that threaten the very existence of your community, you ask yourself, ‘Why did God bless me with this insight?’.

The keen desire to be like the divinely-appointed leaders and growing evidence (in their perception) that they are blessed with capabilities similar to their role models - reinforced by their friends and family over an extended period of time - convinces some religiously devout people that God has chosen them for a special divine role in spreading the message of social identity change urgently needed to save their community from imminent ruin. Believing one to be a divine authority and publicly proclaiming as such are two different things, especially because the very desire for the office makes one ineligible for holding it! Thus, it is not very surprising that religious movement founders are so reluctant to proclaim their religious authority. In 1901, Ahmad claimed that God had told him prior to 1881 that he was a prophet but he had been reluctant to accept such a heavy load (Ahmad, 1901). The question is: What is it that made him overcome his reluctance and publicly proclaim what he had known in his heart for at least seven long years? What is it that makes religious movement leaders overcome their reluctance and make public their divine appointments to prophetic offices?

There is little doubt that friends and family play a crucial role in the formation of our perceptions of who we are. Ahmad only publicly proclaimed his prophetic role once he was surrounded by people who were supportive of his exceptional capabilities. Despite his 1881 revelations from God that he was the prophet, Ahmad refused – until 1889 - repeated requests by Nur-ud-Din and other devotees to accept their pledges of allegiance. After his Sunni wife and sons refused to support his claims, he separated from them and remarried in 1884. His second wife, Nusrat Jehan Begum and her staunchly Ahl-e-Hadis family were fully supportive of Ahmad’s claims. It took another five years of that supportive environment before Ahmad’s sense of prophethood became so fully developed that he started accepting bayah - oaths of allegiance - from his eager followers.

With their friends and family’s support, the sense that they have to take up the role of divinely-appointed messenger to convey the urgently-needed message of social identity change to their community, becomes strong enough for some religious leaders to start publicly proclaiming their perceived role in life. This is the second step in Colliers & Collero's model where the religious innovators start announcing to the world that "I am a divinely sanctioned religious authority."

Those claimants who are lucky attract enough friends and family members to start a new community. The devotees who accept the claimants as messengers of God look to them as model citizens whose words are to be closely followed and whose actions are to be emulated. For these devotees the claimants become the very embodiment of what it means to be a messenger of God. The entire life narrative of the claimant, every action they are seen to take, and every utterance they are heard to utter is justified by the devotees as righteous prophet-like behavior. The claimant's high status becomes part of devotees' social identity and any slight against the claimant is seen as a doctrinal objection against the group's religious ideology. Thus, the claimants are seen as defending the group rather than themselves when they defend their high status against the objections of their detractors. These leaders must publicly integrate their beliefs and behaviors with those associated with their adopted role. This is the third and final stage of the Colliers & Collero's model and indicates a complete adoption of a role by an individual.

The social-identity-change-entrepreneurship model is a general socio-cognitive framework which specifies that some staunch group members concerned that unless their group changes some of its shared religious beliefs its future is bleak, with the help of their family and friends, convince themselves and some devotees that they are divinely-sanctioned religious authorities. However, the general framework does not imply that different religious movements and/or their founders are interchangeable. This is because the social-identity problems faced by various social, cultural, and religious groups are unique and the solutions derived by the social-identity-change-entrepreneurs must also be distinct. For instance, the problems that nineteenth century North Indian Sunni Muslims were facing were unique in many ways. Unlike Muslims in Southeastern Europe, the Middle East, and North Africa, who had lived with Christians right from the beginning of Islam in these areas, Indian Muslims had never encountered large numbers of Christian missionaries and/or lived next to large practicing Christian communities. Prior to the nineteenth century, Hindus had been the neighbours and primary outgroup for North Indian Muslims. But with the British takeover of India, they not only had to live under the rule of a Christian monarch but they also had to deal with thousands of aggressive Christian missionaries successfully employing sophisticated evangelizing techniques to try to convert them. This was a unique challenge that required a unique solution custom-designed by Mirza Ghulam Ahmad. This simply could not have happened anywhere else or at any other time.

Endnotes

1. Moulana (Arabic for "our lord") is a title used for traditional Islamic scholars and clerics and Shaheed is a title used to refer to martyrs. My father's proper name Ali Haidar (given by the second Ahmadiyya Khalifa in whose reign my father was born) was unusual for a non-shia since it is a combination of the two names of Prophet Muhammad's cousin and son-in-law Ali. He was awarded the title *Shaheed* in a Friday sermon by the fourth Ahmadiyya Khalifa because of my father's death in a car accident while carrying out his duties as an Ahmadiyya Muslim missionary in April 1995.
2. Also known as Ahmadiyya Muslim Jama'at (Jama'at is Arabic word for community or movement) and sometimes Ahmadiyya Movement in Islam in the West.
3. *Khalifa-tur-rasul* (Arabic word for successor to the prophet) is the traditional title assigned by Sunni Muslims to the leaders who headed the Muslim community following the death of Prophet Muhammad in 632 AD. Ahmadis extend that title to the successors to Mirza Ghulam Ahmad, the founder of Ahmadiyya Movement.
4. This ban led to the famous incident where Pakistan's Ahmadi foreign minister Sir Muhammad Zafrullah Khan watched the funeral prayers (instead of taking part in them) of Pakistan's founder and Governor General Muhammad Ali Jinnah because they were led by a non-Ahmadi.
5. The commong saying among Ahmadis is that "missionary vows" not only bless the life of the pledger but are also passed on to seven generations of his progeny.
6. They had little option other than supporting PPP since the Islamic parties did not allow Ahmadis to become members because they were considered heretical for believing in a prophet after prophet Muhammad.
7. We lived in Tahrik-e-Jadid Quarter Number 52 from 1983 to 1990. Too old and feeble to take care of himself, Abbaji, my maternal grandfather, moved with us to the Missionary Quarters.
8. This included winning the top prize in the international Ahmadiyya Men's speech and religious knowledge competitions held at what turned out to be *the very last* International *Khuddam-ul-Ahmadiyya Markaziya* (International Ahmadiyya Men's Association) competition held in 1989. Because of bans by government of Pakistan, the International Khuddam Competitions have not been held since 1989, thus I can technically claim to be the reigning Ahmadiyya religious knowledge and Urdu speech champion!
9. The Arabic word Jama'at can be translated as party or community.
10. This also means that until Jama'at reinstates international Khuddam organizations and annual competitions, my status as the reigning Ahmadiyya religious knowledge and Urdu speech champion will remain safe!
11. Mirza Ghulam Ahmad's exact date of birth cannot be established with accuracy. The modern official position of the Ahmadiyya Jama'at is that Ahmad was born

on 13 February 1835 but it has been suggested by Ahmad himself as well as many critics that he was born some time in 1839.

12. I do not use the term Qadiani in a prejoritive sense used by the anti-Ahmadiyya community to refer to all Ahmadis. Here I simply use "Qadiani Ahmadis" to distinguish them from members of the Anjuman Ishat-Islam Lahore.
13. There are no authoritative numbers on the total number of Ahmadis around the world. Jama'at's own numbers have ranged from 70 million to 200 million. Non-Ahmadi estimates range from 10 to 20 million. Some of the difference may be due to various ways of counting (only regular due paying members versus the total number of those who have ever had any affiliation with the Jama'at at some point in their lives) but still the 200 million number seems rather exaggerated.
14. A companion of Syed Ahmad Barelvi. Ismael was one of Syed Ahmad's 600 militiamen who fought and died in the battle of Balakot against the Sikhs in 1831 (see Chapter 4 for more on Syed Ahmad).
15. "The Apostle of Allah (peace be upon him) went out on the day of Badr along with three hundred and fifteen (men). The Apostle of Allah (peace be upon him) said: O Allah, they are on foot, provide mount for them; O Allah, they are naked, clothe them; O Allah, they are hungry, provide food for them." Sunan Abu Dawud: Book 14, Number 2741.
16. When referring to nodes corresponding to mental concepts represented in a spreading activation semantic memory I will use upper case letters unless the context makes it clear that I am referring to mental concepts.
17. This was my way of paying homage to Walter Kintsch who used the term in (Kintsch & Green 1978). Curiously, the term also appears in the book "Made to Stick" by Heath and Heath published in 2007 (two full years after I presented my work at the 2005 Cognitive Science Society Conference where Chip Heath also presented a paper) without any reference to my work, to Kintsch or to anyone else!
18. Readers do frequently (and for the most part successfully) infer genres by reading text even when such information is not obvious and use this information to reason about the text. For more see (Zwaan 1994).
19. Merriam-Webster defines fable as, "a fictitious narrative or statement: as a: a legendary story of supernatural happenings b: a narration intended to enforce a useful truth".
20. The context includes the mental knowledge that a reader has in her memory when processing a concept.
21. Indeed an idea may be counterintuitive for a person in one context but not counterintuitive in another context.
22. From July 1 to July 9, 2010, Angus Reid Public Opinion conducted an online survey among 1,009 Canadian adults who are Angus Reid Forum panellists, 1,002 American adults who are Springboard America panellists, and 2,011 British adults who are Springboard UK panellists. The margin of error—which measures

sampling variability—is +/- 3.1% for Canada and the United States, and 2.2 per cent for Great Britain.

23. The online poll of 2,126 U.S. adults was conducted between November 10 and 17, 2008 by Harris Interactive.
24. *Hazoor* is the honorary title used by Ahmadis to refer to a holy person such as the Prophet Muhammad, one of his four Caliphs, Promised Messiah or one of his Caliphs.
25. Only a small minorities of Ahmadis ever became moosies. An overwhelming majority chose not to become members of the wasiyyat system. As of 2010, out of millions of people who had accepted Ahmadiyyat in the 121 history of the movement's existence, only about ten thousand had chosen to become moosies.
26. As we will see in the next chapter, this metaphor was also a favourite of Mirza Ghulam Ahmad.
27. Mirza Ghulam Ahmad's exact date of birth cannot be established with accuracy. The modern official position of the Ahmadiyya Jama'at is that Ahmad was born on 13 February 1835 but it has been suggested by Ahmad himself as well as many critics that he was born in 1839.
28. Farukh Qadiani was the nom de guerre used by Mirza Ghulam Ahmad in his early poetry
29. There are no authoritative numbers on the total number of Ahmadis around the world. Jama'at's own numbers have ranged from 70 million to 200 million. Non-Ahmadi estimates range from 10 to 20 million. Some of the difference may be due to various ways of counting (only regular due paying members versus the total number of those who have had any affiliation with the Jama'at at some point in their lives) but still the 200 million number seems exaggerated.
30. He was dragged to court by the local mullah for having posted a sign outside the building saying "Ahmadiyya Muslim Mission." Luckily, President Zia-ul-Haq hadn't yet imposed his draconian anti-Ahmadiyya ordinance prohibiting Ahmadis from "pretending to be Muslim" otherwise he surely would have ended up in jail with dozens of other Ahmadis who were prosecuted and sentenced following the promulgation of the ordinance.
31. The Lahori account of the split is specified in Ali (1918) and the Qadiani account can be found in: Ahmad (1924).
32. A century later Pakistan's military dictator Zia-ul-Haq amended the very article to make blasphemy a capital crime in Pakistan. Ironically enough, a number of Pakistani Ahmadis have been prosecuted under the amended 298 for blaspheming prophet Muhammad by accepting Ahmad as a prophet after him!
33. A number of studies (Upal 2007; Heath & Luwenstein 2005; Upal 2013) have documented the use of surprise in advertising.

Glossary of Arabic/Urdu Terms

Adab: Urdu for etiquettes and usually describes etiquettes associated with court and poetry recital culture.
Ahadith: Plural of *hadith* (saying of the prophet Muhammad) collected by Muslims starting about two centuries after his death.
Ahl-e-hadith: A South Asian branch of Sunni Islam that accepts both Quran and Hadith as scriptures but downplays the authority of the four traditional schools of jurisprudence namely, Hanafi, Maliki, Shafii, and Hanbali. It was launched in the nineteenth century by a group of North Indian Muslims spearheaded by Syed Nazir Hussain, Muhaddis Dehlavi, and his disciples including Nawab Sidiq Hasan Khan of Bhopal and Muhammad Hussain Batalavi.
Ahl-e-Quran: A branch of Sunni Islam that that accepts the Quran as the scripture but downplays the authority of Hadith.
Allah: The proper name of God in Islam.
Alim: Arabic for scholar, traditionally reserved for a scholar of Islam trained in a religious madrassa. Plural *ulema.*
Arya Samaj: (Sanskrit for noble society) is a Punjabi Hindu reform movement launched by Swami Dyananda in 1875 that focused on Vedic studies.
Ashraf: The term used by upper-class Indian Muslims to describe themselves.
Bahishti Maqbara: Heavenly cemetery reserved for Ahmadis who pay their tithe and fulfil other conditions laid out by Ahmed in his book, "The Will."
Bayat/Bayah: Oath of allegiance typically offered to a sufi spiritual guide (a pir, murshid, or shaikh) by a *murid* i.e., student.
Daras: A religious lecture usually delivered following a ritual namaz prayer by the imam. In Ahmadiyya mosques, this is usually done after the afternoon (asar) prayer during the month of Ramadhan and after the evening (maghrib) prayer during the remaing 11 months of the year.
Ghair-muqalid: Arabic for a non-follower (opposite of *muqalid*). Usually refers to a Sunni Muslim who does not follow one of the four schools of jurisprudence.
Ghair-Muqalideen: Arabic for non-followers (opposite of Muqalideen). Usually refers to Sunni Muslims who do not follow one of the four schools of jurisprudence. Plural of *ghair-muqalid.*
Hadith: Saying of the prophet Muhammad collected by Muslims starting about two centuries after his death. Sunni and Shia Hadith collections differ as do the perceptions of authenticity of various collections within each tradition.
Hazoor: The honorable one. Usually used by Muslims as a pronoun to refer to a holy person such as the prophet Muhammad. Ahmadis also use it to refer to Ahmad and his Khalifas.
Ijtehad: The authority of an Islamic scholar to carry out independent reasoning in situations where ruling by a traditional authority does not exist.

Ijtema: Annual gatherings of Ahmadiyya auxiliary organizations (separate organizations for boys, girls, adult-men, women, and senior-men).
Imam: The prayer leader who leads Muslims in the communal ritual prayers.
Imam Mahdi: A Muslim eschatological character who along with a returned Jesus will play a critical role in global acceptance of Islam before the end of the world.
Jalsa Salana: The annual 3-day Ahmadiyya Jama'at Convention instituted by Ahmad in 1891 as an opportunity for the faithful to refresh their faith. Originally, it was only held in the Ahmadiyya headquarters f Qadian and Rabwah on the last weekend in December but now the gatherings are held in dozens of countries around the world at all times of the year.
Kafir: Arabic term for those who reject. Commonly used to refer to those who do not believe in Islam.
Khanidan: Urdu word for family. Ahmadis use it to refers to the progeny of Mirza Ghulam Ahmad.
Kuffar: Plural of *Kafir*. Infidels. Those who do not believe in Islam.
Masjid: Arabic word for mosque. A place where Muslims gather for their ritual prayers (called *namaz* in Urdu/Persian and *salat* in Arabic).
Moulavi: An Islamic scholar who is usually also an Imam (i.e., a prayer leader).
Muhaddis: A scholar of Hadith who specializes in understanding authenticity of ahadith through their chains of narration.
Mughals: A group of central Asian warriors who ruled much of Northern India from thirteenth to seventeenth century.
Muqalid: Someone who practices *taqlid* i.e., follows an Islamic scholar's rulings on jurisprudence.
Muqalideen: Plural of *Muqalid* meaning followers.
Murid: A student of Sufism. A murid follows his pir or murshid's guidance to learn Sufism.
Murshid: A spiritual guide and Sufi teacher who has reached the end of the path.
Murtad: Arabic word for apostate. A Muslim who has no longer believes in the truth of Islam.
Namaz: Persian term for a ritual Islamic prayer (known as *salat* in Arabic) including the five daily obligatory prayers of *fajr* (pre-dawn), *zuhr* (early afternoon), *asar* (later afternoon), *maghrib* (evening), and *isha* (night).
Persianate Society: A culture that uses Persian as a language for its high culture.
Pir: A Sufi teacher.
Qibla: The direction in which Muslims offer their ritual prayers (i.e., the direction to Kaba in Mecca).
Quran: The scripture revealed to prophet Muhammad that forms the holy scripture of Islam.
Ramadhan/Ramazan: The ninth month in the lunar Islamic calendar (called Hijri calendar) which is considered to be the holy month of fasting during which Quran was revealed to prophet Muhammad.

Seerat: Biography. Usually refers to biography of a holy person such as the prophet Muhammad.
Shaheed: Arabic word for martyr.
Shaikh: A Sufi teacher who is the head of a spiritual order.
Sayyids: People who claim to be decedents of the Prophet Muhammad through his daughter Fatima.
Shia Islam: Shiaism is a branch of Islam that believes in the right of prophet Muhammad's progeny through his daughter Fatima and son-in-law Ali to hold the leadership of the Islamic community. They believe that Abu-bakr, Umar, and Usman were not rightful leaders and that Ali was the true Imam after prophet Muhammad's death.
Sunni Islam: Sunniism is a branch of Islam that do not believe that prophet Muhammad's progeny are the only ones qualified to hold the leadership of the community. They believes in the four Caliphs Abu-bakar, Umar, Usman, and Ali as rightful Khalifas.
Tahajjud: The optional ritual Islamic prayer (i.e., in addition to the five obligatory ritual prayers) that can only be offered at pre-dawn hours.
Taqlid: Arabic word for follow. Usually refers to the practice of following a *mujtahid* i.e., a scholar qualified to rule on Islamic jurisprudence.
Tariqa: A Sufi spiritual order such as Qadiri, Chishti, and Suharwardi.
Ulema: (plural of *alim*) Arabic word meaning scholars, traditionally reserved for scholars of Islam trained in a religious madrassa.
Wahabis: An Islamic reform movement launched by eighteenth century Najdi Arab preacher, Muhammad Abdul Wahab. It is considered to be the dominant belief system in Saudi Arabia.

Bibliography

Addison, James Thayer (1929). "The Ahmadiyya movement and its western propaganda", *Harvard Theological Review*, 22(1), 1-29.

Ahmad, Basharat (1939). *Mujaddid-e-Azam: Hissa Awal, az ibtida ta June 1900*. Lahore: Ahmadiyya Anjuman Ishat-e-Islam.

Ahmad, Basharat (1924). *Truth About the Split*. Islam International Publications, Qadian.

Ahmad, Basharat (1925). *Jadeed Ahmadiyya Nizam-e-Amal*. Qadian: Zia-ul-Islam Press.

Ahmad, Mirza Bashir (1935). *Seerat-ul-Mahdi*. Qadian, India: Muhtamim Ahmadiyya Kitab Ghar.

Ahmad, Mirza Ghulam (1884). *Braheen-e-Ahmadiyya (Vol. 4)*. Amritsar: Riaz-e-Hind Press.

Ahmad, Mirza Ghulam (1891a). *Azala Aouham*. Amritsar, India: Riaz-e-Hind Press.

Ahmad, Mirza Ghulam (1891b). *Fath-e-Islam*. Amritsar, India: Riaz-e-Hind Press.

Ahmad, Mirza Ghulam (1891c). *Tauzih-e-Maram*, Amritsar, India: Riaz-e-Hind Press.

Ahmad, Mirza Ghulam (1894). *Atmam-ul-Hujjate Alalazi Lija wa Zagh Anil Majat*, Lahore: Gulzar Muhammad Press.

Ahmad, Mirza Ghulam (1897). *Anjam-e-Atham*. Qadian, India: Zia-ul-Islam Press.

Ahmad, Mirza Ghulam (1898). *Kitab-ul-Bariya ma Ayat Rab al Bariya*. Qadian, India: Zia-ul-Islam Press.

Ahmad, Mirza Ghulam (1902). *Tiryak-ul-Quloob*. Amritsar, India: Zia-ul-Islam Press.

Ahmad, Mirza Ghulam (1903). *Daily Al-Badr Qadian*, 2(5), 36-37.

Ahmad, Mirza Ghulam (1905a). *Al-Wasiyat*. Qadian: Zia-ul-Islam Press.

Ahmad, Mirza Ghulam (1905b). *Lecture Ludhiana*. Qadian, India: Zia-ul-Islam Press.

Ahmad, Mirza Ghulam (1906). *Chashma-e-Masihi*. Qadian, India: Zia-ul-Islam Press.

Ahmad, Mirza Ghulam (1907). *Tatimma Haqeeqat-ul-Wahi*. Qadian, India: Matba Magazine.

Ahmad, Mirza Ghulam (1908). *Maseeh Hindustan Main*. Qadian, India: Matba Anwar-e-Ahmadiyya Machine Press.

Ahmad, Mirza Ghulam (2005). *The Will*, Tilford, UK: Islam International Publications. Translation of (Ahmad 1905a).

Ahmad, Mirza Ghulam (2007). *Fountain of Christianity*. Surrey, UK: Islam International Publications. Translation of (Ahmad, 1906).

Ahmad, Mirza Ghulam (2008). *Maktoobat-e-Ahmad Volume 1 & 2* (S. A. Hai & Y. A. Irfani, Eds). Tilford, UK: Islam International.

Ahmad, Shazia (2010). "Proselytizing through Print: The Ahmadiyya from 1880-1908". Paper presented at *the Workshop for Identifying New Themes in South Asian History*, Zentrum Moderner Orient (ZMO) in Berlin.

Ahmadiyya-Gazette (2013). "Khalifa of of Islam makes historical address to European Parliament". *Ahmadiyya Gazette Canada*, January 2013 issue. http://www.ahmadiyyagazette.ca/Articles/khalfa-of-islm-makes-historic-address-at-european-parliament.html retrieved on 1 January 2017.

Akyol, Mustafa (2011). *Islam without extremes: A Muslim case for liberty*. New York: WW Norton.

Ali, Muhammad (1917). *Hazrat Mirza Ghulam Ahmad and the Finality of Prophethood*. Lahore: Anjuman Ishat-e-Islam.

Ali, Muhammad (1918). *The Ahmadiyya Movement IV: The Split*. Lahore: Coloured Printing Press.

Ali, Muhammad (1937). *The Founder of Ahmadiyya Movement*. Lahore: Ahmadiyya Anjuman Islam.

Alladin, Hafiz M. S. (2015). "The Sign of Heavens". *The Review of Religions*, August 2015.

Allen, Charles (2006). *God's terrorists: The Wahabi cult and the hidden roots of modern Jihad*. Cambridge, UK: Da Capo Press.

Amritsari, Sanaullah (2002). *Radd-e-Qadiyaniat*. Multan, Pakistan: Aalami Majlis Tahaffuz Khatam-e-Nabuwwat.

An-Nadawi, S. A.-H. A. (1974). *Qadi'ianism: A critical Study*. Lucknow, India: Academy of Islamic Research and Publications.
Anderson, John R. (2005). *Cognitive Psychology and Its Implications*, 6th Edition, New York, NY: Worth Publishers.
Anderson, John R. (1983). "A spreading activation theory of memory". *Journal of Verbal Learning and Verbal Behavior*, 22.
Atran, Scott (2004). *In Gods We Trust: Evolutionary Landscape of Religion*. Oxford, MA: Oxford University Press.
Attia-Krieger, Sharon (2014). *Sacrifice Scripts: The role of context in the transmission of counter-cultural religious representations of sacrifice and commitment. Israeli-Jewish culture*. PhD Thesis, London School of Economics.
Bainbridge, William Sims (1985). "Cultural Genetics". In Stark, Rodney (ed.) *Religious Movements: Genesis, Exodus, and Numbers* (pp. 157-198), New York, NY: Paragon House.
Bainbridge, William Sims, & Stark, Rodney (1979). "Of Churches, Sects, and Cults: Preliminary Concepts for a Theory of Religious Movements". *Journal for the Scientific Study of Religion*, 18(2), 117-133.
Bar-Tal, Daniel (2000). *Shared Beliefs in a Society: Social psychological analysis*. Thousand Oaks, CA: Sage.
Baron, Jonathan (1998). *Judgement Misguided: Intuition and Error in Public Decision Making*. New York, NY: Oxford University Press.
Barrett, Justin L., & Nyhof, Melanie (2001). "Spreading non-natural concepts: The role of intuitive conceptual structures in memory and transmission of cultural materials". *Cognition and Culture*, 1, 69-100.
Barrett, Justin L. (1997). *Anthrpomorphism, intentional agents, and conceptualizing God*. PhD Thesis, Cornell University, Ithaca, NY.
Barrett, Justin L. (2000). "Exploring the Natural Foundations of Religion". *Trends in Cognitive Sciences*, 4, 29-34.
Barrett, Justin L. (2008). "Coding and Quantifying Counterintuitiveness in religious concepts: Theoretical and methodological reflections". *Method and Theory in the Study of Religion*, 20, 308-338.
Barrett, Justin L., & Keil, Frank C. (1996). "Anthropomorphism and God concepts: Conceptualizing a non-natural entity". *Cognitive Psychology*, 31(3), 219-247.
Batalavi, Muhammad Hussain (1884). *Ishat-us-Sunnah*, 8(3), 158-348.
Bauldelaire, Eric, & Lynch, Dov (2009). *Documentaire: Abkhazia Journal*. Retrieved 1 May 2009, from http://documentaire.com/caucasus/Abkhazia.html
Bebbington, David W. (1989). *Evangelicalism in Modern Britain: A History from the 1730s to the 1980s*. New York: Routledge.
Beecher, Henry W. (1887). *A summer in Englad with Henry Ward Beecher*. London, UK: Ford, Howard, & Hulbert.
Beveridge, Albert J. (1900). *Congressional Record, Senate, 56th Congress, 1st session, Jan. 9, 1900*, 704–712.
Bigelow, Anna (2005). "Punjab's Muslims: The history and significance of Malerkotla". *Journal of Punjab Studies*, 12(1), 63-95.
Bloch, Maurice (2005). "Are religious beliefs counter-intuitive". In M. Bloch (Ed.), *Essays on Cultural Transmission* (pp. 103-123). New York: Berg.
Bloom, Paul A., & Fischler, Ira (1980). "Completion norms for 329 sentence contexts". *Memory and Cognition*, 8, 631-642.
Bower, Gordon H., Black, John B., & Turner, Timothy J. (1979). "Scripts in Comprehension and Memory". *Cognitive Psychology*, 11, 177-220.
Boyer, Pascal (1994). *The Naturalness of Religious Ideas: A Cognitive Theory of Religion*. Berkeley, CA: University of California Press.

Boyer, Pascal (2001). *Religion Explained: The evolutionary origins of religious thought.* New York, NY: Basic Books.

Boyer, Pascal, & Ramble, Charles (2001). "Cognitive templates for religious concepts". *Cognitive Science*, 25, 535-564.

Bruner, Jerome (1991). "The narrative construction of reality". *Critical Inquiry*, 18(1), 1-21.

Bukhari, Imam Muhammad (1997). *The Translation of the Meanings of Sahih Al-Bukhari: Arabic-English.* Trans. Muhammad Ahmad Khan. Houston, TX: Dar-us-Salam Publications.

Calgary Herald (2008)." Largest Mosque in Canada Opened: Harper, Dion both on hand to see sacrifices pay off". *Calgary Herald*, 6 July 2008 issue.

Campo, Juan E. (2009). *Encyclopedia of Islam*. New York, NY: Infobase Publishing.

Carson, Penelope S. (1988). *Soldiers of Christ: Evangelicals and India, 1784-1833.* PhD Thesis, King's College, London.

Centola, Damon, Willer, Robb, & Macy, Michael (2005). "The emperor's dilemma: A computational model of self-enforcing norms". *American Journal of Sociology*, 110(4), 1009-1040.

Chryssides, George D. (2011). *Historical Dictionary of New Religious Movements* (Second ed.): Scarecrow Press.

Clark, Henry Martin (1907). *Robert Clark of the Punjab: Pioneer and Missionary Statesman.* London, UK: Andrew Melrose.

Clark, Robert (1883). *A Brief Account of Thirty Years of the Missionary Work of Church Missionary Society in the Punjab and Sindh 1852 to 1882.* Lahore: Albert Press.

Collier, Peter J. & Callero, Peter L. (2005). "Role theory and social cognition: learning to think like a recycler". *Self Identity*, 4 (1), 45–58.

Committee-on-Inaugural-Ceremonies (1989). *Inaugural Addresses of the Presidents of the United States: From George Washington, 1789 to George H. W. Bush, 1989*, New York, NY: Cosimo Books.

Compilation-Committee (1863). *Report of the Punjab Missionary Conference (held at Lodhiana in December/January 1862-63)*, Lahore: Committee of Compilation.

Converse, Philip E. (1964). "The nature of belief systems in mass publics". In D. E. Apter (Ed.), *Ideology and Discontent* (pp. 206-261). New York: Free Press.

Cox, Jeffrey (2002). *Imperial Fault Lines: Christianity and Colonial Power in India, 1880-1940.* Stanford, CA: Stanford University Press.

Crawford, Paul, & Brown, Brian (2002). "'Like a friend going round': reducing the stigma attached to mental healthcare in rural communities". *Health and Social Care in the Community*, 10(4), 229-238.

Dard, Abdul Rahman (1948). *Life of Ahmad: Founder of Ahmadiyya Movement.* Lahore, Pakistan: Tabshir Publications.

Davidson, Denise, Larson, Shari, Luo, Zupei, & Burden, Matthew (2000). "Interruption and bizarreness effects in the recall of script based text". *Memory*, 8(4), 217-234.

Dawud, Abu (2000). *Sunan Abu Dawud: English translation.* Trans. A. Hasan. New Delhi, India: Kitab Bhavan.

Dennett, Daniel C. (2006). *Breaking the spell: Religion as a natural phenomenon.* New York: Penguin.

Dupont, Jerry (2001). *The Common Law Abroad: Constitutional and Legal Legacy of the British Empire.* Littleton, CO: Fred B Rothman and Co.

East India Proclamations (1908). *Copies of the Proclamation of the Kind, Emperor of India, to the Princes and Peoples of India, of the 2nd day of November 1908, and the Proclamation of the late Queen Victoria of the 1st day of November 1858, to the Princes, Chiefs, and People of India*, London, UK: Eyre and Spotttiswoode Ltd.

Ellis, Nick C. (2006). Language acquisition as relational contingency learning. *Applied Linguistics*, 27(1), 1-24.

Fabiani, Monica, & Donchin, Emanuel (1995). "Encoding processes and memory organization: A model of the von Restorff effect". *Journal of Experimental Psychology: Learning, Memory, & Cognition*, 21, 224-240.
Festinger, Leon, Riecken, Henry, & Schachter, Stanley (1964). *When Prophecy Fails*. New York, NY: Harper Torch Books.
Feurstein, George (1991). *Holy Madness: The Shock Tactics and Radical Teachings of Crazy-Wise Adepts, Holy Fools, and Rascal Gurus*. New York, NY: Paragon.
Freud, Sigmund (1964). *The Future of an Illusion*. New York, NY: Doubleday.
Friedman, Yohanan (1992). *Prophecy Continuous: Aspects of Ahmadi Religious Thought and Its Medieval Background*. New Delhi, India: Oxford University Press.
Furnish, Timothy R. (2005). *Holiest Wars: Islamic Mahdis, their jihads, and Osama bin Laden*. Westport, CT: Praeger Publishers.
Fusfeld, Warren (1992). "Review of Prophecy Continuous". *International Journal of Middle East Studies*, 24(2), 347-348.
Gardner, Martin (1957). *Fads and Fallacies in the Name of Science*. New York: Dover.
Gigerenzer, Gerd, Todd, Peter, & ABC Research Group (1999). *Simple Heuristics That Make Us Smart*. New York: Oxford University Press.
Gill, Paul (2007). "A Multi-Dimensional Approach to Suicide Bombing". *International Journal of Conflict and Violence*, 1(2), 142-159.
Gonce, Lauren, Upal, M. Afzal, Slone, D. Jason, & Tweney, Ryan (2006). "Role of Context in the Recall of Counterintuitive Concepts". *Cognition and Culture*, 6(3-4), 521-547.
Gopnik, Allison, & Meltzoff, Andrew N. (1997). *Words, Thoughts, and Theories*. Cambridge, MA: The MIT Press.
Grant, Charles (1813). *Observations on the State of Society among the Asiatic Subjects of Great Britain, particularly with respect to Morals, and on the means of Improving It. Written chiefly in the Year 1792*. London: House of Commons.
Griffin, Lepel Henry, & Massy, Charles Francis (1890). *The Panjab Chiefs: Historical and biographical notices of the principal families in the Lahore and Rawalpindi divisions of the Panjab (Vol. 2)*. Lahore: Civil and Military Gazette Press.
Griswold, Eliza (2010). *The Tenth Parallel: Dispatches from the fault line between Christianity and Islam*. new York.
Griswold, Hervey De Witt (1902). *Mirza Ghulam Ahmad: The Mehdi Messiah of Qadian*. Lodhiana, India: American Tract Society.
Gruss, Edmond C., & Thuet, Lane A. (1977). *What Every Mormon (And Non-mormon) Should Know: Examining Mormom History, Doctrine and Claims*. New York, NY: Xulon Press.
Guenther, Alan M. (1998). *Controversy as a "Necessary Evil"? Perspectives on Missions to Muslims in India in the late 19th century*. Cambridge, UK: NAMP Research Forum, Westminster College.
Guiliani, Rudolph W. (1994). *Inauguration Speech by Rudolph W. Guiliani*, New York, NY: The City of New York, Retrieved 4 may 2015, from http://www.nyc.gov/html/records/rwg/html/96/inaug.html
Guthrie, John T., Wigfield, Allan, Metsala, Jamie L., & Cox, Kathleen E. (1999). "Motivation and Cognitive predictors of text comprehension and reading amount". *Scientific Studies of Reading*, 3, 231-256.
Harmon-Vukic, Mary & Upal, M. Afzal (2012). "Understanding the Memory Advantage of Counterintuitive Concepts". *Journal of Religion, Brain & Behavior*, 2 (1), 121-139.
Harmon-Vukic, Mary, Upal, M. Afzal, Sheehan, Kathy (2011). "The Role of Mental Elaborations in Making Counterintuitive Ideas Memorable". In *Proceedings of the 33rd Annual Meeting of Cognitive Science Society* (pp. 985-990), Mahwah: Lawrence Erlbaum.
Haslam, Alexander, Reicher, Stephen, & Hopkins, Nick (2005). "Social identity and the dynamics of leadership: Leaders and followers as collaborative agents in the transformation of social reality". *Leadership Quarterly*, 16, 547-568.

Haven, Kendall, & Ducey, Mary G. (2007). *Crash Course in Storytelling (Vol. 388)*, Westport, CT: Libraries Unlimited.
HC (2010). *UK House of Commons Hansard*, London, UK: House of Commons, URL: https://hansard.parliament.uk retrieved 1 July 2015.
Heath, Chip, & Heath, Dan (2007). *Made to Stick: Why some ideas survive and others die*. New York, NY: Random House.
Heit, Evan (1998). "Influence of prior knowledge on selective weighting of category members". *Journal of Experimental Psychology: Learning, Memory, and Cognition*, 24, 712-731.
Hirt, E. R., & Clarkson, J. J. (2010). "The psychology of fandom: Understanding the etiology, motives and implications of fanship". In L. Kahle & A. Close (Eds.), *Consumer Behavior Knowledge for Effective Sports and Event Marketing*. New York, NY: Routledge.
Hockney, Mike (2014). *Why Math Must Replace Science*, Miami, FL: Hyperreality Books.
Hodgson, Marshall (1974). *The Venture of Islam: The expansion of Islam in the Middle Periods*. Chicago, IL: University of Chicago Press.
Hogg, Michael A., & Vaughan, Graham M. (2002). *Social Psychology* (3rd ed.). London, UK: Prentice Hall.
Hughes, T. P. (1875). "Notes on Muhammadanism, being Outlines of the Religious System of Islam". London: W. H. Allen.
Hughes, Thomas P. (1879). "Church Missionary Society Peshwar Missionary Report". In Conference-Secreataries (Ed.), *Proceedings of the General Conference On Foreign Missions* (pp. 323-333). London, UK: John F. Shaw & Company.
Hunt, R. Reed, & Worthen, James B. (2006). *Distinctiveness and Memory*. New York, NY: Oxford University Press.
Hussain, Syed Shahzad (1998) "I was tortured by 'Muslims' but I refused to recant". In Ahmad, Nurudin, & Malik, Naveed (Eds.) *By the Dawn's Early Light: Short Stories by American Converts to Islam* (pp. 122-125), Silver Spring, MD: Majlis Khuddam ul Ahmadiyya US.
Iannaccone, Laurence R. (1998). "Introduction to economics of religion". *Journal of Economic Literature*, 36, 1465-1496.
Ikram, Sheikh Muhammad (2003). *Mauj-e-Kausar*, Lahore, Pakistan: Jadeed Press.
Imad-ud-din. (1893a). "Dr. Imad-ud-din's paper for Chicago". *The Church Missionary Intelligencer: A Monthly Journal of Missionary Information*, 44, 579-584.
Imad-ud-din. (1893b). "Christian Efforts Among Indian Mohammedans: Being an account of the effects of the teaching of the Bible amongst the Mohammedans of India, together with a consideration of the question how many of them have become Christians and why". *The Church Missionary Intelligencer*, 44, 584-590.
Ishtihar Qeemat Nur Afshan. (1873). *Nur Afshan*, Vol. 1, p. 1.
Jeyaraj, D. (2008). "Mission reports from South India and their impact on the Western mind: The Tranquebar mission of the eighteenth century". In D. L. Robert (Ed.), *Converting Colonialism: Visions and Realities in Mission History*, 1706-1914 (pp. 21-42). Grand Rapids, MI: William. B. Eerdmans Publishing Co.
Jost, John T. (2006). The end of the end of ideology. *American Psychologist*, 61, 651-670.
Jukes, W. (1925). *Reminiscences of missionary work in Amritsar 1872-1873 and on the Afghan Frontier in Peshawar 1873-1890: Project Canterbury Archives*. Retrieved 12 April 2011, from http://www.AnglicanHistory.org
Just for Laughs (2015a), *Just for Laughs Gags*, URL: https://www.youtube.com/user/JustForLaughsTV. Retrieved on 1 April 2015.
Just for Laughs (2015b), *Sheep Dog Prank*, URL: https://www.youtube.com/watch?v=fnxtbUwRU3c&t=10s, Retrieved on 1 April 2015.
Just-for-Laughs, (2015c) *Purchase Tester Casher Gag*, URL: https://www.youtube.com/watch?v=6H5L576A-Uw, Retrieved on 1 April 2015.

Keil, Frank C. (1981). "Constraints on knowledge and cognitive development". *Psychological Review*, 88, 197-227.
Khaldun, Ibn (1967). *The Muqaddimah: An introduction to history.* Trans. F. Rosenthal. Princeton, NJ: Princeton University Press.
Khan, Adil Hussain (2015). *From Sufism to Ahmadiyya: A Muslim minority movement in South Asia.* Indianapolis, IN: Indiana University Press.
Khan, Syed Ahmad (1880). *Tafsir-e-Ahmadi*, Aligarh, India: Aligrah Institute Press.
Khan, Syed Ahmad (1885). *The Life and Works of Sir Syed Ahmed Khan CSI*, London, UK: William Blackwood and Sons.
Khan, Syed Ahmad (1972). *Writings and Speeches of Sir Syed Ahmad Khan*, Mohammad, S. (Translator). Bombay, India: Nachiketa Publications
Khan, Syed Ahmad (1872). *Review on Dr. Hunter's Indian Musalmans: Are they bound in conscience to rebel against the Queen?* Benaras, India: Medical Hall Press.
Khan, Syed Ahmad (1959). *Maktoobat-e-Sir Syed (Vol 1)*, Panipati, S. M. I. (Ed.) Aligarh, India: Majlist Taraqi-e-Adab.
Kintsch, Walter (1998). *Comprehension: A paradigm for cognition.* Cambridge, MA: Cambridge University Press.
Knapp, B. (1991). *The Destiny of the Mother Church*, Christian Science Publication Society.
Lavan, Spencer (1974). *The Ahmadiyya Movement: A history and perspective.* New Delhi, India: Manohar Book Services.
Lawson, E. Thomas, & McCauley, Robert N. (1990). *Rethinking Religion: Connecting cognition and culture.* Cambridge, UK: Cambridge University Press.
Lawson, Todd (2009). *The Crucifixion and the Qur'an: A Study in the History of Muslim Thought.* New York: Oneworld Publications.
LeVine, Robert A., & Cambell, Donald T. (1972). *Ethnocentrism: Theories of conflict, ethnic attitudes, and group behavior.* New York: Wiley.
Lewis, James (2008). *Oxford Handbook of New Religious Movements*, Oxford, UK: Oxford University Press.
Longley, Clifford (2003). *Chosen People: The Big Idea that Shaped England and America.* London, UK: Hodder & Stoughton.
Lugli, U. (2013). "The Concept of Myth". *Journal of Studies in Social Sciences*, Vol 6(1), 38-57.
Marable, Manning & Agard-Jones, Vanessa (2008). *Transnational Blackness: Navigating the Global Color Line*, New York, NY: Palgrave MacMillan.
Masson, Charles (1842). *Narrative of Various Journeys in Balochistan, Afghanistan and the Panjab.* London: Richard Bentley.
Mayer, T. J. Lee (1876). From Letters of Rev. T. J. Lee Mayer. *The Church Missionary Intelligencer And Record: A Monthly Journal of Missionary Information*, 1, 427-428.
McCauley, Robert N., & Lawson, E. Thomas (2002). *Bringing Ritual to Mind: Psychological Foundations of Cultural Forms.* Cambridge, UK: Cambridge University Press.
McDaniel, Mark A., Einstein, Giles O., DeLosh, Edward L., & Brady, Paul (1995). "The bizzarreness effect: It's not surprising, it's complex". *Journal of Experimental Psychology: Learning, Memory, and Cognition*, 22, 422-435.
McKee, Scot (2002). *Creative B2 Branding (no really).* New York, NY: Longfellow Publishers.
McLellan, Jaime L. (2012). *Theology of Revolution: Messianic Traditions and the Revolutions They Inspired.* MA Thesis, University of Kansas.
Melton, Gordon (2008). "An Introduction to new religions". In J. R. Lewis (Ed.), *The Oxford Handbook of New Religious Movements* (pp. 16-35). Oxford, UK: Oxford University Press.
Melton, Gordon, & Baumann, Martin (2010). *Religions of the World: A comprehensive encyclopedia of beliefs and practices.* Santa Barbara, CA: ABC-CLIO.

Metcalfe, Barbara D. (1982). *Islamic Revival in British India: Deoband, 1860-1900*. Princeton, NJ: Princeton University Press.
Metcalfe, Barbara D. (2009). *A Historical Overview of Islam in South Asia*. In B. Metcalfe (Ed.), Islam in South Asia in Practice. Princeton, NJ: Princeton University Press.
Mummendey, Amelie, & Schreiber, Hans-Joachim (1984). "'Different' just means 'better': Some obvious and some hidden pathways to in-group favouritism". *British Journal of Social Psychology*, 23, 363-368.
Murphy, Gregory L., & Medin, Douglas L. (1985). "The role of theories in conceptual coherence". *Psychological Review*, 92(3), 289-316.
Myer, David E., & Schvaneveldt, Roger W. (1971). "Facilitation in recognizing pairs of words: Evidence of dependence between retrieval operations". *Journal of Experimental Psychology*, 90, 227-234.
Myers, Jerome L., O'Brien, Edward J., Balota, David A., & Toyofuku, Maria (1984). "Memory search without interference: The role of integration". *Cognitive Psychology*, 16, 217-242.
Naeem, Nadhra Shahbaz (2010). "Life at the Lahore Darbār: 1799-1839". *South Asian Studies*, 25(2), 283-301.
Naim, Choudhri M. (2001). "Ghalib's Delhi: A shamelessly revisionist look at two popular metaphors. Paper presented at *the Urdu Scholarship in Transnational Perspectives*". Columba University. http://www.urdustudies.com/pdf/18/06NaimGhalib.pdf
Nawaz, Maajid (2012). *Radical: My Journey out of Islamist Extremism*. London, UK: Virgin Books.
Nesgoda, Kevin (2010). *Eagles are Easily the best team in the NFC after the win over Giants*. Retrieved 22 November 2010, from http://bleacherreport.com/articles/523897-eagles-are-easily-the-best-team-in-the-nfc-now-after-win-over-giants
Newberg, Andrew, & D'aquili, Eugene (2008). *Why God Won't Go Away: Brain Science and the Biology of Belief*. New York, NY: Random House.
Newell, Alan (1990). *Unified Theories of Cognition*. New York, NY: Harvard University Press.
Nizami, Farhan A. (1983). *Madrasahs, Scholars and Saints: Muslim Responses to the British Presence in Dehli and the Upper Doab 1803-1857*. D. Phil. Thesis, Oxford University, Oxford, UK.
Norris, Pippa, & Inglehart, Ronald (2011). *Sacred and secular: Religion and politics worldwide* (2nd Ed.). New York: Academic Press.
Norton, Sean & Upal, M. Afzal (2014). "Narratives, Identity, and terrorism". In Bramadat, Paul & Dawson, Lorne (eds.) *Religious Radicalization & Beyond* (pp. 285-300), Toronto: University of Toronto Press.
Notovitch, Nicolas (1990/1887). *The Unknown Life of Christ*, San Diego, CA: Tree of Life Publishing.
Numani, Sirajulhaq (1915). *Tazkiratul Mahdi*. Qadian, India: Zia-ul-Islam Press.
Nur Afshan (1873). Shifa'at kay bayan mein. (10 July 1873). *Nur Afshan*, published from Lodhiana, India by Presbyterian Mission in the Punjab.
Nur Afshan (1875). Shifa'at kay bayan mein. (3 June 1875). *Nur Afshan*, published from Lodhiana, India by Presbyterian Mission in the Punjab.
Nur-ud-Din, Hakeem (1908, 22 April 1908). *Al-Hakm*, published from Qadian, India by Yaqub Ali Irfani.
Packer, Dominic (2008). "On Being Both With Us and Against Us: A Normative Conflict Model of Dissent in Social Groups". *Journal of Personality and Social Psychology*, 12(1), 50-72.
Pazzani, Michael (1991). "The influence of prior knowledge on concept acquisition: Experimental and computational results". *Journal of Experimental Psychology: Learning, Memory, and Cognition*, 17(3), 416-432.
Pew-Research-Center (2013). *The World's Muslims: Religion, Politics and Society*, Washington, DC: Pew Research Center.
Philips, Jonathan (2015). "The Crusades: A Complete History". *History Today*, 65(5), May 2015, URL: http://www.historytoday.com/jonathan-phillips/crusades-complete-history, retrieved on 1 March 2017.

Porter, Andrew N. (2000). “Evangelicalism, Islam, and Millennial Expectation in the Nineteenth Century”. *International Bulletin of Mission Research*, 24(3), 111-118.

Powell, Avril Ann (1993). *Muslims and Missionaries in Pre-Mutiny India*. Abingdon, UK: Routledge Press.

Quran (2000). *The Holy Quran*. Trans. Ali, Abdullah Yusuf. Hertfordshire, UK: Wordsworth.

Rahman, Fazlur (1958). “Muslim Modernism”. *Bulletin of the School of Oriental and African Studies*, 21(1), 82-99.

Rasmussen Report (2010). *73% agree that Washington is broken*. Retrieved 31 December 2010, from http://www.rasmussenreports.com/public_content/politics/general_politics/february_2010/73_agree_that_washington_is_broken

Reicher, Stephen, Haslam, Alexander, & Platow, Michael J. (2011). *The new psychology of leadership: Identity, influence and power*. New York: Psychology Press.

Reynolds, Gabriel S. (2009). “The Muslim Jesus: Dead or alive?” *Bulletin of the School of Oriental and African Studies*, 72(2), 237-258.

Richardson, Valerie (2015). “Muslim Reform Movement Decries Radical Islam, Calls for Equality”. *Washington Times*, December 6, 2015, URL: http://www.washingtontimes.com/news/2015/dec/6/muslim-reform-movement-decries-radical-islam-calls/, retrieved on 6 March 2017.

Riesmann, Catherine K. (2001). “Analysis of Personal Narratives”. In Gubrium, Jaber & Holstein, James (Eds.), *A Handbook of Interview Research*. Thousand Oaks, CA: Sage Publications.

Robinson, Francis (1988). *Varieties of South Asian Islam*. Warwick, UK: Centre for Research in Ethnic Relations, University of Warwick.

Robinson, Francis (2000). *Islam and Muslim History in South Asia*. Delhi, India: Oxford University Press.

Robinson, Francis (2002). “The Ummah in 21st century”. *The Herald Weekly Karachi*, 108.

Russell, Yvan I., & Gobet, Fernand (2013). “What is counterintuitive? Religious cognition and natural expectation”. *Review of Philosophy and Psychology*, 4(3), 715-749.

Saliba, John (2004). “Psychology and new religious movements”. In J. R. Lewis (Ed.), *The Oxford Handbook of New Religious Movements*. New York, NY: Oxford University Press.

Schank, Roger C. (1999). *Dynamic Memory Revisited*. New York, NY: Cambridge University Press.

Schooler, Lael J. (1993). *Memory and the Statistical Structure of the Environment*. PhD Thesis, Carnegie Mellon University, Pittsburgh, PA.

Schooler, Lael J., & Anderson, John R. (1997). “The role of process in the rational analysis of memory”. *Cognitive Psychology*, 32(3), 219-250.

Shahid, Dost Muhammad (1958). *Tareekh-e-Ahmadiyyat*. Rabwah: Idarat-ul-Musanifeen.

Sheehy-Skeffington, Jennifer (2008). “Extending the counterintuitive effect: Implicit theories of Social behavior and the cognitive impact of the counter cultural”. In *Proceedings of the 31st Annual Scientific Meeting* (pp. 9-12), Paris, France: International Society for Political Psychology.

Shupe, Anson D., & Bromley, David G. (1981). “Apostates and atrocity stories: Some parameters the dynamics of deprogramming”. In B. R. Wilson (Ed.), *The Social Impact of New Religious Movements*. New York: Rose of Sharon Press.

Sikand, Yoginder (2005). *Bastion of the Believers: Madrasas and Islamic Education in India*. New Delhi, India: Penguin.

Silverstein, Adam & Stroumsa, Guy (2015). *The Oxford Handbook of the Abrahamic Religions*, Oxford, UK: Oxford University Press.

Simon, Herbert, & Newell, Alan (1972). *Human Problem Solving*. Englewood Cliffs, NJ: Prentice Hall.

Singh, Ganda (Ed.). (1952). *The Punjab in 1839-40: Selections from the Punjab Akhbars, Punjab Intelligence, etc. Preserved in the National Archives of India, New Delhi*. Patiala, India: New Age Press.

Singh, Khushwant (1963). *A History of the Sikhs: Volume 1 1469-1839*, Princeton, NJ: Princeton University Press.

Singh, Teja (2006). *A Short History of the Sikhs (Vol. 1)*. Patiala, India: Punjabi University.

Siraj-ud-din (1913). "The vital forces of Christianity and Islam". *International Review of Mission*, 2(1), 96-117.

Slone, D. Jason (2002). *Theological Incorrectness: Why Religious People Believe What They Shouldn't: Explaining Theological Incorrectness in South Asia and America*. PhD Thesis, Department of Comparative Religion, Western Michigan University, Kalamazoo, MI.

Slone, Jason, Gonce, Lauren, Upal, M. Afzal, Edwards, Kritin, Tweney, Ryan (2007). Imagery effects on recall of minimally counterintuitive concepts, *Cognition and Culture*, 7(3-4), 355-367.

Smith, Vincent Arthur (1923). *The Oxford History of India: From the earliest times to the end of 1911* (2nd ed.). Oxford, UK: Clarendon Press.

Sperber, Dan (1996). *Explaining culture: A naturalistic approach*. Malden, MA: Blackwell Publishers.

St. John, Warren (2005). *Rammer Jammer Yellow Hammer: A Road Trip Into the Heart of Fan Mania*. New York: Three Rivers Press.

Stark, Rodney, & Bainbridge, William Sims (1987). *A Theory of Religion*. New York, NY: Peter Lang.

Storr, Anthony (1996). *Feet of clay: Saints, sinners, and Madmen. A Study of Gurus*. New York, NY: Free Press.

Suvorova, Anna (2004). *Muslim Saints of South Asia: The eleventh to fifteenth century*. Oxon, UK: Routledge Curzon.

Tajfel, Henri, & Turner, John (1979). "An Integrative Theory of Intergroup Conflict". In Austin, W. & Worchel, S. (Eds.), *The Social Psychology of Intergroup Relations*. San Francisco, CA: Brooks-Cole.

Tajfel, Henri, & Turner, John (1986). "The social identity theory of intergroup behavior". In Worchel, S. & Austin, W. (Eds.), *Psychology of Intergroup Relations*. Chicago, IL: Nelson Hall.

Turner, John, Hogg, Michael A., Oakes, Penelope J., Reicher, Stephen, & Wetherell, Margaret (1987). *Rediscovering the social groups: A self-categorization theory*. Oxford, UK: Basil Blackwell.

Tweney, Ryan, Upal, M. Afzal, Gonce, Lauren, Slone, D. Jason, & Edwards, Kristin (2006). "The Creative Structuring of Counterintuitive Worlds". *Cognition and Culture*, 6(3-4), 483-498.

Upal, M. Afzal (2005a). "Role of Context in Memorability of Intuitive and Counterintuitive Concepts". In Bara, Bruno G., Barsalou, Lawrence & Bucciarelli, Monica (Eds.), *Proceedings of the 27th Annual Meeting of the Cognitive Science Society*, 2224-2229, Mahwah, NJ: Lawrence Earlbaum.

Upal, M. Afzal (2005b). "Towards A Cognitive Science of New Religious Movements". *Cognition and Culture*, 5(2), 214-239.

Upal, M. Afzal (2007a). "The Optimal Cognitive Template of Minimally Counterintuitive Narratives". In McNamara, D. S. & Trafton, J. G. (Eds.) *Proceedings of the 29th Annual Meeting of the Cognitive Science Society*, Mahwah, NJ: Lawrence Erlbaum.

Upal, M. Afzal (2007b). "What is More Memorable Counterintuitive Concepts Interpreted Metaphorically or Literally?" In McNamara, D. S. & Trafton, J. G. (Eds.), *Proceedings of the 29th Annual Meeting of the Cognitive Science Society*, Mahwah, NJ: Lawrence Erlbaum.

Upal, M. Afzal (2009). "An Alternative Account of the Minimal Counterintuitiveness Effect". *Journal of Cognitive Systems Research*, 11(2), 194-2003.

Upal, M. Afzal (2011a). "From Individual to Social Counterintuitiveness: How layers of innovation weave together to form tapestries of human cultures". *Mind and Society*, 10(1), 79-96.

Upal, M. Afzal (2011b). "Memory, Mystery and Coherence: Does the presence of 2-3 counterintuitive concepts predict cultural success of a narrative?" *Journal of Cognition and Culture*, 11(1-2), 23-48.

Upal, M. Afzal (2014a). "A Framework for Agent-based Social Simulations of Social Identity Dynamics. In P. von Fellman, Y. Bar-Yam & A. A. Minai (Eds.), *Countering Terrorism, Insurgency, Ethnic and Regional Violence* (pp. 89-109). Berlin: Springer Verlag.

Upal, M. Afzal (2014b). “Three practical lessons from the science of influence operations design”. *Canadian Military Journal*, 14(2), 53-58.

Upal, M. Afzal (2014c). “Cognitive Framework for counterintuitive story understanding”. In Finlayson, Mark Meister, C. & Bruneau, E. (Eds.) *Proceedings of the Fifth International Workshop on Computational Modeling of Narratives* (pp. 224-242), Berlin: Schloss Dagstuhl.

Upal, M. Afzal (2015a). “Alternative Narratives for Preventing the Radicalization of Muslim Youth”. *Journal of Deradicalization*, 15(2), 138-162.

Upal, M. Afzal (2015b). “Countering Islamic Jihadist Movements”. *Journal of Terrorism Research*, 6(2), 57-69.

Upal, M. Afzal (2015c). “Why Stickiness is not enough to explain persistence of counterintuitive religious concepts”. In Noelle, D. C., Dale, R., Warlaumont, A. S., Yoshimi, J., Matlock, T., Jennings, C. D., & Maglio, P. P. (Eds.) *Proceedings of the 37th Annual Meeting of the Cognitive Science Society* (pp. 2452-2456), Austin, TX: Cognitive Science Society.

Upal, M. Afzal (2015d). “Why Moderate Muslims Balk at ‘Je Suis Charlie’”. *Jihad Watch*, 2 February 2015, https://www.jihadwatch.org/2015/02/why-moderate-muslims-balk-at-je-suis-charlie, retrieved on 3 February 2015.

Upal, M. Afzal, & Gibbon, S. (2015). “Agent-based simulation of social identity dynamics”. In Saikou, Diallo, Navonil, Mustafee, Mittal, Saurabh, & Jafer, Shfagh (eds.) *Proceedings of the Spring Simulation Multi-Conference* (pp. 185-192), San Diego, CA: The Society for Modeling & Simulation International.

Upal, M. Afzal, Gonce, Luren, Tweney, Ryan, & Slone, D. Jason (2007). “Contextualizing counterintuitiveness: How context affects comprehension and memorability of counterintuitive concepts”. *Cognitive Science*, 31(3), 415-439.

Upal, M. Afzal, Packer, Dominic, Moskowitz, Gordon, Kugler, Brian (2011). *Investigating the Dynamics of Identity Formation, and Narrative Information Comprehension: Final Report*, TR 2011-061, Defence Research & Development Canada-Toronto Research Centre.

van Evera, Stephen (1994). “Hypotheses on Nationalism and War”. *International Security*, 18(4), 5-39.

van Knippenberg, A. (1978). “Status differences, comparative relevance, and intergroup differentiation”. In H. Tajfel (Ed.), *Differentiation between social groups: Studies in the social psychology of intergroup relations* (pp. 171-199). London: Academic Press.

Von Bulow, Bernhard (1915). *Imperial Germany*, Lewenz, Marie (translator), New York, NY: Dodd Mead.

von Restorff, Hedwig (1933). “Analyse von Vorgangen in Spurenfield. I. Uber die Wirkung von Bereichtsbildungen im Spurenfeld” (Analysis of Processes in the Memory Trace). *Psychologishe Forschung*, 18, 299-342.

Walter, Howard Arnold (1918). *The Ahmadiya Movement*. New York, NY: Oxford University Press.

Wilson, Colin (2000). *The Devil’s Party: A History of Charlatan Messiahs*. London, UK: Virgin.

Winter, Andrew (2010). *One Million Celebrate Spanish Homecoming*. Retrieved 25 July 2010, from http://news.sky.com/skynews/Home/World-News/Spain-Win-World-Cup-Spanish-Fans-Welcome-Squad-Home-As-Plane-Lands-In-Madrid-After-Victory/Article/201007215663618

Yamamoto, J. Isamu (1977). *The Puppet Master: An Inquiry Into Sun Myung Moon and the Unification Church*. New York, NY: Inter Varsity Press.

Zwaan, Rolf A. (1994). “Effect of genre expectations on text comprehension”. *Journal of Experimental Psychology: Learning, Memory, and Cognition*, 20, 920-933.

Zweig, Stefan (1932). *Mental Healers (Franz Anton Mesmer, Mary Baker Eddy, Sigmund Freud)*. New York, NY: Viking.

Subject Index

Person Index

Geographic Index